Charles Giles Bridle Daubeny

Climate

An inquiry into the causes of its differences, and into its influence on vegetable life, comprising the substance of four lectures delivered before the Natural history society, at the museum, Torquay, in February, 1863

Charles Giles Bridle Daubeny

Climate

An inquiry into the causes of its differences, and into its influence on vegetable life, comprising the substance of four lectures delivered before the Natural history society, at the museum, Torquay, in February, 1863

ISBN/EAN: 9783337377298

Printed in Europe, USA, Canada, Australia, Japan

Cover: Foto ©berggeist007 / pixelio.de

More available books at **www.hansebooks.com**

CLIMATE:

AN INQUIRY INTO THE CAUSES OF ITS DIFFERENCES, AND INTO ITS INFLUENCE ON VEGETABLE LIFE.

COMPRISING THE SUBSTANCE OF

Four Lectures

DELIVERED BEFORE

THE NATURAL HISTORY SOCIETY,

AT

THE MUSEUM, TORQUAY, IN FEBRUARY, 1863.

BY

C. DAUBENY, M.D., F.R.S.,

FELLOW OF THE LINNÆAN AND GEOLOGICAL SOCIETIES; HONORARY MEMBER OF THE ROYAL IRISH ACADEMY, OF THE ROYAL AGRICULTURAL, AND OF THE MEDICO-CHIRURGICAL SOCIETIES; FOREIGN ASSOCIATE OF THE ROYAL ACADEMY OF MUNICH; CORRESPONDING ASSOCIATE OF THE GIŒNNAN SOCIETY OF NATURAL HISTORY AT CATANIA; MEMBER OF THE SOCIETIES OF QUEBEC, MONTREAL, PHILADELPHIA, AND BOSTON; OF THE ACADEMY OF GENEVA, ETC.;
PROFESSOR OF BOTANY AND OF RURAL ECONOMY IN THE UNIVERSITY OF OXFORD.

Oxford and London:
JOHN HENRY AND JAMES PARKER.

LONDON: H. G. BOHN,
YORK-STREET, COVENT-GARDEN.

1863.

TO

WILLIAM HENRY TINNEY, ESQ., Q.C.,

MASTER IN CHANCERY, ETC.,

These Lectures,

ORIGINALLY DRAWN UP TO MEET THE REQUIREMENTS

OF A POPULAR AUDIENCE,

AND NOW COMMITTED TO PRINT,

IN ACCORDANCE WITH HIS FRIENDLY WISHES

AND HIS INDULGENT ESTIMATE OF THEIR SHORTCOMINGS,

ARE INSCRIBED,

AS A TOKEN OF REGARD AND ESTEEM,

BY

THE AUTHOR.

LIST OF SUBSCRIBERS

UP TO APRIL 6TH.

A. Baldry, Esq., Bythorn, Bronshill Road
A. Barton, Esq., Falkenstein, Cary Road
Dr. Becker, Park Crescent, 4 copies
J. S. Beckett, Esq., Knoll, Barton Road
Mrs. Belfield, Parkfield, Paignton
Rev. H. Biddulph, Carclew, Hesketh Road, 2 copies
W. J. Booth, Esq., Lisworney
Hon. J. Boyle, Rockwood, Park Hill Road
J. Buckton, Esq., Lauriston Hall
Mrs. Burgess, Lisburn Crescent

Miss Burdett Coutts, Ehrenberg Hall, 2 copies
Miss Cole, Lauriston Hall
Dr. Coates, Hillside, Lower Woodfield Road
J. Corrie, Esq., Springfield, Springfield Road
W. H. Cosway, Esq., Fonthill, Lower Warberry Road, 2 copies
— Calhoun, Esq., Madeira Villa, Tor Church Road.
Miss M. Croome, Iffley, near Oxford
A. Cunninghame, Esq., Belgrave House, 2 copies

The Lady Dunsany, Rockland, Cary Road
A. Dendy, Esq., Rock House, Rock Road
G. D. Wingfield Digby, Esq., Sherborne Castle, Dorset
Mrs. Wingfield Digby, ditto
J. Dugmore, Esq., Beacon Terrace
Miss Dyott, Bay Mount, Heathfield, Warberry Road, 2 copies
Mrs. Dykes, Kilmorie, Hesketh Crescent

N. B. Edmonstone, Esq., St. Anne's, Meadfoot Road
Mrs. C. English, Vomero, Stitchill Road, 2 copies
J. Enys, Esq., Enys, near Penryn, 4 copies
Dr. Evanson, Homehurst, Lower Warberry Road

E. S. Ffarington, Esq., Wellswood Park
Rev. R. Fayle, Park Hill Villa, Park Hill Road
J. H. Fenton, Esq., Lauriston Hall

W. Gott, Esq., Bay Fort, Warren Road, 2 copies
Rev. H. Griffin, Glenthorne, Lower Warberry Road

The Viscountess Hood, Barton Seagrave, Kettering
Dr. Radclyffe Hall, Plymswood House, Bronshill Road, 2 copies
Dr. C. Henry, Haffield, near Ledbury, 2 copies
Rev. J. R. Hogg, Eastholme, Upper Lincombe Road

Rev. C. M. Jarvis, Capo di Monte, Lower Warberry Road, 2 copies
Mrs. Jarvis, Brock Street, Bath
Rev. L. Jenyns, Darlington Place, Bath, 2 copies

Sir John Kennaway, Bart., Escot, near Ottery St. Mary, 2 copies
W. H. Kitson, Esq., Vaughan Parade
Sir Richard Kirby, C.B., Lauriston Hall, 2 copies

The Lady Julia Lockwood, Barcombe, Paignton
Captain Vaughan Lee, Osborne House
Rev. S. Lovett, Warren Hall, St. Luke's Road
Major Lumley, Lauriston Hall

Dr. Madden, Gorton, Lower Woodfield Road
W. Marshall, Esq., M.P., Ringwood, Meadfoot, 2 copies

Miss Master, Bay Mount, Park Hill Road
Miss E. M. Mansfield, Birstall, Upper Braddon Hill Road
W. Metcalfe, Esq., Woodleigh Vale, Warberry Road, 2 copies
Miss Caroline Milnes, Frystone, Meadfoot Road
E. Mount, Esq., Modena Terrace

Dr. Nankivell, Layton House, Waldon Terrace
Rev. Dr. Newman, Underheath, Warberry Road
W. Norris, Esq., Belvoir, Meadfoot Road

W. Pengelly, Esq., Lamorna, St. Mary-Church Road
March Phillipps, Esq., Wellswood, Lower Warberry Road
Dr. Lovell Phillips, Torville, Teignmouth Road
Mrs. Potts, Daison, Warberry Road
J. Pynsent, Esq., Linden, St. Luke's Road

W. Rashleigh, Esq., Port Neptune, Fowey
Mrs. G. Richards, Iffley, near Oxford
Mrs. Roberts, Torwood Mount

N. W. Senior, Esq., Kensington Gore
N. G. Senior, Esq., ditto
Miss Senior, ditto
A. B. Sheppard, Esq., Hove, Furze Hill Road
Philip Sleeman, Esq., South Town House
Mad. Smirnoff, Enfield, Meadfoot Road
Rev. W. J. Smithwick, Beacon Terrace
W. D. Splatt, Esq., Abbotsford, Lower Warberry Road
W. Stabb, Esq., Portland Place
J. Stoddart, Esq., Collingwood
Cortlandt Taylor, Esq., Western Terrace, Belgrave Road
Mrs. Taylor, Warberry Lodge, 2 copies
Dr. Tetley, Belmont, Teignmouth Road, 8 copies
Captain Tibbets, Barton Seagrave, Kettering

R. Tighe, Esq., Merton Lodge, Middle Lincombe Road
W. H. Tinney, Esq., Snowdenham, 5 copies
Mrs. Tinney, ditto, 2 copies
Mrs. Alexander Toogood, Gorphwysfa, Warren Road
B. Toogood, Esq., Annandale, Torwood Gardens
Lieut.-Col. H. Trevelyan, Oke Lodge, Southampton
A. Turnbull, Esq., Priory, Park Hill Road

E. Vivian, Esq., Woodfield, Lower Woodfield Road

Miss Walker, Broadlands, Bronshill Road
Rev. Sandys Wall, Beacon Terrace, 4 copies
Rev. G. Warner, Highstead, Bronshill Road
Miss Warrington, Percy Lodge, Abbey Road, 2 copies
J. North White, Esq., St. Hilary, Warren Road
Sherlock Willis, Esq., Torwood Mount
N. Wing, Esq., 2, Park Crescent
Rev. R. R. Wolfe, Furze Park, Furze Hill Road

Dr. Yonge, Plymouth
Hon. Mrs. Yorke, Kanescombe, Lower Warberry Road, 4 copies

Count von Zech, Lisburn Crescent

SUPPLEMENTARY LIST OF SUBSCRIBERS,

TO MAY 22ND.

Sir Thomas Acland, Bart., Killerton, 2 copies.
Thomas Acland, Esq., Exeter, 2 copies.
Rev. P. Arnold, Council Office, Downing-street, London.
Sir B. Brodie, Bart., Oxford.
Rev. Dr. Bulley, Magdalen College, Oxford, 2 copies.
J. Bulley, Esq., Magdalen College, Oxford.
J. Burdon, Esq., Marine Villa, Torquay.
Rev. E. Daubeny, Ampney, Cirencester, 5 copies.
P. Duncan, Esq., Bath.
Lady Easthope, 2, Great Cumberland-street, Hyde Park.
R. Fox, Esq., Falmouth, 2 copies.
Rev. R. Greswell, Worcester College, Oxford.
G. Griffiths, Esq., Jesus College, Oxford.
J. Hoyte, Esq., Glastonbury.
Rev. H. Jenkins, Stanway, Colchester, 4 copies.
Sir C. Lemon, Bart., Carclew.
A. Luscombe, Esq., Coombe Royal, Kingsbridge.
Dr. Masters, Bridgwater.
Sir W. Palmer, Bart., Bridport, 2 copies.
J. Payne, Esq., Magdalen College, Oxford.
Professor Phillips, Oxford.
Professor B. Price, Oxford.
Professor Rolleston, Oxford.
Professor Goldwin Smith, Oxford.

Rev. R. H. Tiddeman, Oxford.
Mr. Weeks, Chemist, Torquay.
Rev. H. Winwood, Bath.
H. Wyndham, Esq., Oriel College, Oxford.

ERRATA.

For Hon. J. Boyle, Rockwood, Park Hill-road, *read* J. Bogle, Esq., Woodside.

To J. Corrie, Esq., *add* 4 copies.

To J. Stoddart, Esq., *add* 3 copies.

For Mrs. Roberts, *read* Robertson.

For E. Mount, Esq., *read* E. Mevert.

LECTURE I.

Introductory remarks. Definition of Climate. Temperature—how far independent of solar influence. Different sources of Heat considered;—all except the solar influence may be passed over. Solar Heat considered. Conjectures as to its nature founded upon the Spectrum analysis. Brief statement of the recent discoveries made on that subject. Methods of determining mean temperatures—new instruments for the purpose. Rate of the decrease of Heat dependent on latitude. Concurrent causes affecting temperature. Instances of places where the actual temperature exceeds the normal one. Instances of the reverse. Tabular view of the temperature of certain places. Summer temperature that which most operates upon vegetation. Causes affecting temperature—general and local. General ones—influence of solar light in tropical countries upon land and water—by day—by night. Influence of solar light in northern latitudes—in summer upon land—upon water. Ditto in winter on each. Speculations as to the existence of a polar sea. Temperature of America and Europe different—and why. Causes of the greater heat of the Globe in former times. Sir C. Lyell's hypothesis—how far admissible. Probable greater preponderance of water over land during former periods of the earth's history.

LADIES AND GENTLEMEN,

IT has been my practice for the last five winters to resort to some mild spot, either in the south of England or on the Continent, with the view of escaping the trying effects of the cold and damp of Oxford upon a chest rather susceptible of such influences; and as I have been compelled during the periods of my absence from the University to intermit the ordinary routine of my duties and occupations, it is a satisfaction to me, whenever an opportunity seems to present itself of labouring in ever so slight a degree in the same direction, by imparting any information that I possess on matters of science, to those who may be thrown in my way in the places to which I have been induced to migrate.

And in the present instance I was also glad to be able to do something towards expressing my acknowledgments for the compliment paid me during my former visit to Torquay by your Natural History Society, in electing me an honorary member of that body; so that when it was intimated by your Secretary and by other influential members that it would be gratifying to them if I were to deliver one or more lectures at their Institution during my stay in the neighbourhood, I could not do less than respond to the call, without waiting to enquire whether the wares, that I had brought with me from one of the supposed great emporiums of learning and science, were of sufficient value to be worth exporting to so great a distance.

Nor had I much difficulty in fixing upon a theme for my discourse—one, that is, which persons of every description could enter into, and which the inhabitants of such a place as this might be expected to regard with especial favour; of a character sufficiently popular to arrest the attention of the many, and yet connected with the highest speculations which can occupy the man of science; in its principles soaring to the sublimest regions of philosophy, in its applications throwing light upon the most important questions which concern the interests of daily life. Need I say that this subject is the Weather, or, in more grandiloquent phraseology, the Science of Meteorology—a science, indeed, still in its infancy, yet even in its infantine state already assuming gigantic proportions; one into which the untutored peasant sometimes would seem to possess an intuitive insight, whilst the philosopher, although he may plume himself on his acquaintance with the general laws of atmospheric phenomena, is often at a loss to unravel the entangled skein of effects connected with it which daily observation brings before him.

Indeed, whilst many a Charlatan, utterly ignorant of science, preys upon the credulity of the public by boldly prognosticating the atmospheric changes that are about to take place, the man of science, who has spent his life in investigating the laws of Nature, will engage with diffidence in any such undertaking; and hence we find Ignoramuses, who figure in print under the sobriquets of Zadkiel, or Thomas Moore, Physician, predicting the weather for a year to come with the most entire confidence, whilst an Herschel or an Arago declare themselves incompetent to anticipate what may chance to supervene within the space of the next four-and-twenty hours.

Not that there is any reason to suppose the changes in the atmosphere less brought about through the operation of secondary forces than other natural phenomena; or to be more immediately under the direction of the great Cause of all than physical events in general are admitted to be; and hence we can perceive no greater inconsistency, when the devout philosopher, after praying for fine weather, speculates upon the meteorological changes which may cause the sun to come out, or the rain to fall, than when a pious general, like an Havelock or a Stonewall Jackson, after supplicating the Almighty for success to his arms, shapes his conduct purely by military considerations as to the question of accepting or declining battle.

Indeed, the most submissive waiter upon providence is, in spite of himself, more or less of a meteorologist.

The first thing which we enquire about in the morning, and the last thing which it occurs to us to speculate upon at night,—especially in so fickle and changeable a climate as our own,—is the weather.

It controls all our proceedings, and modifies all our results; it reads us a constant lesson of humility, by shewing how large a part of the results which we are striving at by the exertion of our own forethought and industry is dependent upon forces over which we have no control; and it affects our calculations, not only throughout all the ordinary transactions of daily life, but when our mind indulges in a wider range, and interests itself respecting the animals, the plants, the statistics, the sanitary condition of other climes and countries; in short, it is of equal concern to the physician, to the painter, to the poet, to the agriculturist, and to the natural historian.

Indeed, the chief reason why my own attention has been drawn to the subject of meteorology, is the influence which Climate exerts upon the plants which characterize different portions of the globe, and upon the methods of cultivating the soil in our own, questions which come before me in my double capacity as Professor of Botany and of Rural Economy.

Nor is it one of the least powerful reasons for introducing a mention of it here, that Torquay, in common with the rest of the coast of Devonshire and Cornwall, presents a meteorological phenomenon almost without parallel in any other part of the globe, namely, as being situated on a line of coast occupying a latitude as high as 50° or 51° N., and yet favoured with a climate so mild, that in various places of Cornwall the myrtle and camellia will grow in the open air to the dimensions of large shrubs, the orange and lemon will bear fruit with little or no protection during the winter, and even in a few of the more favoured spots, the Fan Palm of Europe and the Chusan Palm of China, three species at least of the Dragon-tree, the Pride of India, and the Camphor-tree, may be seen flourishing in the open soil of a garden[a].

Probably the only part of the world which presents an equally striking variation from the normal condition belonging to its latitude is the coast of the Mediterranean, between Marseilles and Genoa, where, between the parallels 43° and 44°, the Date Palm assumes the proportions of a large tree, and many other tropical productions thrive almost as in their native climates.

By the Climate of a country, then, we understand its relations to temperature, light, moisture, winds, atmospheric pressure, electricity, and so forth; but amongst these the first place must be conceded to the intensity of the heat, and to its distribution over different portions of the year, as it is this which in a great degree regulates the other conditions, and is also itself of all others the one most indispensable for the exercise of the functions of vegetable and animal life.

It has been calculated, that were all the existing sources of heat withdrawn, the temperature which the globe possessed would not exceed 76° below zero, or 108° below the point at which water freezes. Fourier indeed assigned to it one considerably higher, namely 46° below zero, or 85° below the freezing point of water, but as a cold of — 76° has been actually observed in the open air at Melville Island, and one of —72° 4', according to Erman, at Yakoutzk in Siberia, it does not seem possible to suppose, that a temperature higher than

[a] See Appendix.

this could prevail over the earth, if the sun were blotted out of the firmament.

Now this temperature, which is after all probably far removed from the point of the absolute negation of heat, and indeed is not so low as has been often produced by artificial means [b], is conjectured to arise from the heat distributed over the globe by the innumerable stellar bodies which emit rays from their orbits.

This, indeed, is a more important element in its influence upon terrestrial bodies than might at first sight be supposed. Fourier remarks, that if the celestial spaces were entirely devoid of heat, the decrease of temperature from the equator to the poles ought to proceed in a much higher ratio than is the case, and also that the cold in high latitudes would be incalculably greater.

The least variation in the distance of the sun would produce extreme differences in the climate of a place, and the transition from day to night would be attended with a much greater change of temperature than is found to occur. It is therefore to be inferred, that a physical cause is always present, which moderates the temperature of the earth's surface, and imparts to it a fundamental heat which is independent of the sun.

Now the excess of heat beyond this point, even in the coldest part of the globe, namely in the Arctic regions, averages probably not less than 76 degrees of Fahrenheit, as the mean temperature of the polar regions is generally set down as about zero of that scale; whilst in the tropics it is about twice as great; so that, curiously enough, it would appear that nearly half as much heat is obtained extraneously in the Arctic regions, as in those at or near the equator.

The sources from which this accession of temperature beyond that of the space in which our planet is moving can be supposed to be derived, are the following: 1st, the heat generated by the various living bodies scattered over the face of the globe; 2nd, that resulting from various processes

---

[b] M. Natterer by mixing liquid protoxide of azote and bisulphide of carbon, and placing the mixture *in vacuo*, produced a cold of $-220°$ F. This is the lowest temperature hitherto obtained. Miller, vol. i. p. 254.

carried on either by natural agencies or through the instrumentality of man over various parts of the globe, including the different forms of combustion; 3rd, the internal heat of the globe; 4th, the radiation from the sun.

It must be admitted, that all animals generate a certain amount of heat, in proportion to the energy with which their vital functions are conducted, and that a powerful local influence is also exerted upon the temperature of particular spots by artificial combustion. But both these agencies, although jointly they may affect to a certain extent the temperature of cities, where large masses of people are congregated, seem wholly inadequate to produce any sensible change in the climate of the globe generally.

It was indeed suggested to me by an ingenious member of this Society, that a certain calorific influence must be assigned to all the great movements of the air and water—to the action of the tides, and to the waves that dash against the coast, &c.; nor, if Grove's views be correct, is it possible to deny, that some effect ought to arise from the conversion of motion into heat, according to the views at present entertained with regard to the transmutation of natural forces one into the other. This, however, it is impossible to estimate, and therefore in the present enumeration of the causes which tend to elevate the general temperature of the earth's surface, it will be necessary to pass them over. Nor does the internal heat of the globe, whatever it may have done in former times, exercise any material influence over the crust at present; so that in considering the subject of climate, this element also may be thrown out of the calculation altogether, as not affecting the result at the present time.

There remains, therefore, only the direct effect of the sun, concerning the nature of which we, as meteorologists, have but little to do.

It may be sufficient to say that, according to Arago, the body of the solar orb is itself almost entirely dark, but is encompassed at a considerable distance by a luminous envelope, in which funnel-shaped openings exist, through which portions of the dark body below are sometimes seen, producing what are called the spots on its surface.

But both below and above this envelope there is, according to the same philosopher, an atmosphere of vapour, the former intervening between the luminosity and the body of the solar orb, the latter surrounding it externally; and from the undulations or protuberances of its surface, producing those red mountain or flame-like forms which are so remarkable in every total eclipse, as in the last one so well photographed by Mr. Delarue in the north of Spain.

But it is not light only, but also heat, which is radiated from the sun; and hence, in spite of the authority of Arago, and the plausible explanation of the spots in the sun which his hypothesis affords, it may be inferred, that the surface of a body contiguous, as the orb of the sun is, to a photosphere or zone of vapour of so exalted a temperature, would be itself rendered incandescent, and therefore luminous. Be that, however, as it may, we are induced, from the property common to all solid bodies, when intensely heated, of becoming luminous, to conclude, that the rays of light emitted from this source must arise from particles of matter brought into a high state of incandescence.

Of what nature these particles of matter may be, a beautiful discovery lately made in Germany affords us, perhaps, the means of conjecturing.

It had been known ever since the time of Newton, that a sunbeam passed through a prism is separated into seven distinct portions, possessed of different colours; and it was pointed out by Wollaston, and afterwards confirmed by Frauenhofer, that in the spectrum produced by thus refracting the sun's rays, a vast number of dark lines exist, which can be distinguished if the spectrum be sufficiently magnified.

Now it has been found, that if any mineral substance be exposed to a heat sufficient to cause it to volatilize, and if the light produced by its incandescence be transmitted through a similar prism, a spectrum is produced which, although in other parts dark, as compared with that from the sun, exhibits, at some one or more particular points, luminous bands of a certain colour, and that both the colour and the position of the band is always dependent upon the nature of the base which enters into the composition of the mineral body employed.

Many of these, indeed, require for their volatilization a higher temperature than can be produced by an ordinary lamp; and these, of course, can only be rendered sufficiently luminous by the more intense heat produced by other means, as, for instance, by electricity. Such are the metals, iron, copper, gold, silver, &c., and their compounds; but others, like sodium, lithium, calcium, strontium, and barium, will exhibit these characteristic bands when introduced into a common gas lamp of Bunsen's construction, when the light produced by the incandescent body is passed through a prism.

Thus the smallest portion of a salt of sodium, such, for instance, as common salt, produces a bright yellow line in that part of the spectrum which is indicated by the letter A; lithium by one red line at A, and one yellow at B; strontium by a broad orange line at A, followed by a number of narrow streaks of fainter red at B; calcium by a broad greenish line at B, followed by a number of narrow streaks of yellow, and terminated by a bright broad orange line at A; barium, by a numerous succession of faint blue and brighter green and yellow bands in the places indicated in the chart [c].

The quantity of each of these bodies capable of producing these bands of colour was found to be inconceivably small, it being calculated by Kirchkoff and Bunsen, that the one hundred and eighty millionth part of a grain of sodium could be rendered apparent by the bright yellow band characteristic of that metal; and such infallible indices are these bands of the presence of the body, and so exactly do they maintain in each instance their relative position in the spectrum, as ascertained by the most rigorous measurements, that when the above philosophers perceived other lines, not coincident with those noticed in the spectra, to be produced upon rendering certain other bodies incandescent by the same method, it was inferred that they

[c] I refer to the chart constructed by Kirchkoff with the view of exhibiting these spectra.

must have arisen from the presence of some other substances not hitherto discovered. And this inference, wonderful to relate, was verified by the detection of two new metals in the mineral water of Durkheim in Bavaria, the residuum of which had been introduced into the flame of the lamp, as in the other cases. One of these bodies is called cæsium, the other rubidium; and both have been separated from the water, and examined by the chemists alluded to, although existing in it in such minute quantities, that in order to obtain 105 grains of the chloride of cæsium, and 135 of that of rubidium, no less than forty tons of the mineral water of Durkheim were evaporated. They have been since detected in the ashes of plants, and seem to be generally distributed, although in minute quantities, throughout nature.

An English chemist, Mr. Crookes, has since been led, by the same mode of investigation, to the discovery of a third substance, called thallium, present in iron pyrites, which though it resembles in appearance lead, approaches in chemical properties more nearly to the alkaline metalloids.

On this new method, however, which has received the name of Spectrum Analysis, my time does not permit me to dwell; but what more relates to our present purpose, is the inference which the discoverers of this mode of research have deduced, as to the nature of the bodies which, by their incandescence, give rise to the sun's light.

It has been found that every body absorbs the description of rays which itself sends forth. Sodium, for instance, which emits yellow light, intercepts that of the same colour or quality, so that by interposing the vapour of this metal in the path of a ray containing yellow light, the portion of the spectrum, in which the latter exists, becomes extinguished, and a dark band takes its place.

If, therefore, a ray proceeding from the incandescent body of the sun passes through the vaporous atmosphere surrounding its orb, any yellow light present in it will be cut off, if sodium be present in a volatile condition in the medium which it traverses, and hence a dark streak will result, exactly corresponding in position and in magnitude to the bright sodium band.

Hence the existence in the solar spectrum of a dark line, exactly corresponding in point of position with the yellow sodium band, is regarded as an evidence that this metal exists in the solar orb, as well as in the photosphere surrounding it, the former emitting, the latter absorbing, the yellow ray; and as this vaporous atmosphere is infinitely less bright than the orb itself, the rays which the former emits are too feeble to be perceived in conjunction with those of the latter, so that a dark band exists in the spectrum where the yellow line of sodium would otherwise have appeared.

Proceeding upon the same principle, we ascribe the other dark lines which exist in the sun's spectrum to the presence of particular metals in its orb, wherever these lines are found to coincide in point of position with the luminous bands produced by these same substances in a spectrum obtained artificially, the dark band in these cases, as in the former one, being attributed to the extinction of the light in that particular position during its passage through a medium in which the substance itself was present.

On these data, then, it is concluded, that iron exists in the solar atmosphere, as the particular bright lines produced by the introduction of this metal into an artificial spectrum are reversed in the solar one, by dark lines precisely corresponding in position to the former.

By a similar method of research, Kirchkoff concludes that the solar atmosphere contains lime, magnesia, and soda. Chromium has likewise been recognised in it by the same means; nickel too appears to be present, as well as barytes, copper, and zinc in small quantities; but of the remaining metals no indications have as yet been obtained, nor, what is remarkable, does silica appear to be present.

Thus by this remarkable discovery, which would indeed surpass our powers of belief, if it had not been verified by the best possible test, namely, by the detection through its means of several new and before unsuspected substances, we seem to have obtained a glimpse of the physical constitution of that great luminary, which is placed at the unapproachable distance of ninety-five millions of miles from the earth we inhabit. Moreover, by employing the same method, it has since been inferred, that the composition of the fixed stars is not always identical with that of the sun; Sirius, for instance, and some others that have been examined, presenting a spectrum differing from that of our own luminary, and one which appears to indicate the existence of other elements [d].

But it is time to terminate this digression, for which my best apology is the great general interest felt in the novel and beautiful investigation alluded to, as it is with the effects of the solar orb upon the condition of our own planet, rather than with its own intrinsic nature, that we as meteorologists have to deal.

These effects, it is evident, will be experienced more sensibly in proportion to the directness or perpendicularity with which its rays impinge upon the earth's surface, so that they will be least energetic in the polar regions, and most so within the tropics.

If indeed the globe presented throughout a solid surface, unchequered by any irregularities, and uniform both as to texture and colour, the temperature of each portion might be calculated by simply appealing to its latitude, since each parallel would differ from the one above or below it in a certain fixed and easily ascertainable ratio.

Were the earth for instance, as the ancients considered it, and as for convenience sake it is represented on Mercator's projection, a plain surface, it is evident that, granting the mean temperature of the equator to be 80°, and that of the poles zero of Fahrenheit, there would be a decrease of $8\frac{1}{3}°$ of Fahr. for every addition of 10° of latitude.

But as it is a sphere, the ratio is different; and it appears

---

[d] Professor Donati, in a recent memoir read before the Astronomical Society, has classified a few of the more conspicuous of the fixed stars into four groups, according to the kind of light they emit, namely, into white, yellow, orange, and red stars, and has observed that whilst in each class the position of the dark lines corresponds one with the other, every separate group differs from the remaining ones, as well as from the solar orb, in that respect. Ch. News, March 21, 1863.

from the calculations of Professor Dove, who is regarded as the highest authority on the subject of temperature, that if the annual mean at the equator be reckoned at 79°, there would be no difference, but rather a slight increase, in the heat of the globe in all the latitudes within 10 degrees of the line; after which, in the 20th parallel, the temperature would sink to 76°; in the 30th, to 68°; in the 40th, to 56°: in the 50th, to 41°; in the 60th, to 30°; in the 70th, to 17°; and that it would stand nearly at zero in the 80th, the highest latitude yet attained by man, from which point to the poles there would probably be found little or no deviation as to temperature.

Thus the decrement of heat from the equator up to the 20th parallel does not exceed 3° of Fahr.; between 20° and 30°, and 30° and 40°, it goes on pretty uniformly at the rate of 8° of Fahr. for 10° of latitude; it then progresses in an increasing ratio from 40° to 50°, namely, about 15° of temperature for 10° of latitude; diminishes again between 50° and 60°, between which parallels the rate of decrease is 11° of temperature for 10° of latitude; rises again to 13° between 60° and 70°, where the difference amounts to 13°; and attains its highest rate of diminution between 70° and 80°, where it is as much as 17°.

The difference between the actual rate of diminution in temperature according to latitude, and that which would prevail if the earth had been a flat surface with the sun stationed exactly perpendicularly to the equator, may be exhibited in a tabular form by a curved line describing the rate of decrease in temperature in different latitudes. See Table opposite.

It is, however, evident, that although the figures I have given represent the heating effect produced by the direct solar influence upon each portion of the globe, so far as it is due to latitude, so many concurrent causes contribute to the aggregate effect, that few spots exactly accord with this theoretical estimate, some falling short of it, and a still greater number exceeding the number assigned.

But before entering upon this subject it will be well to

LATITUDES.

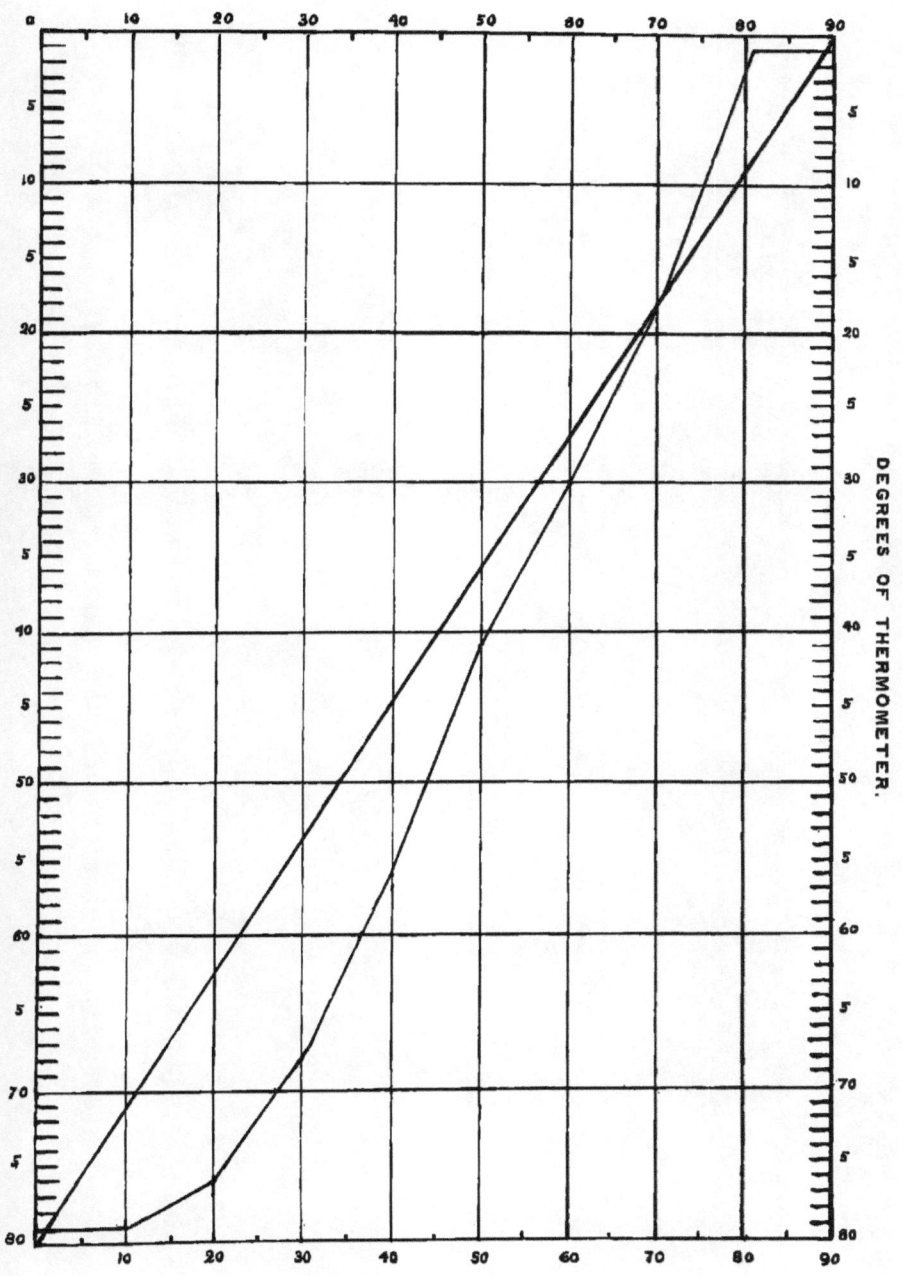

point out the methods by which meteorologists ascertain the mean temperature of different spots on the globe. This is not so simple and easy a problem as might at first be conceived. Two methods are generally proposed for the purpose.

The first is to determine the maximum and minimum temperature during every twenty-four hours, and to assume the mean between the two as the mean of the climate.

The maximum and minimum points are ascertained by means of a self-registering thermometer, either that of Six or that of Rutherford having been generally employed for that purpose.

These answer very tolerably, the former one especially, when the instruments are stationary; but for travelling it is necessary to substitute others.

For ascertaining the maximum temperature, one which is easily managed, not liable to get out of order in being conveyed from place to place, and at the same time sufficiently exact, is one, the principle of which we owe to my friend and colleague, Professor Phillips of Oxford.

In this, it will be perceived by the annexed drawing, that the thread of mercury being broken, the detached portion is pushed forwards as the temperature advances, and remains at the point which it had reached, when the subsequent decrease of temperature causes the remainder of the column to recede.

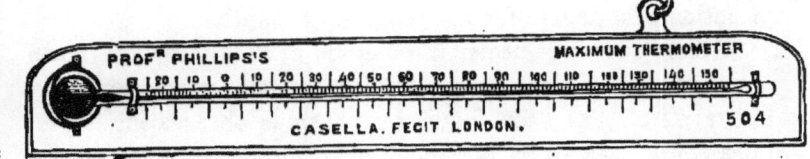

The most convenient and accurate thermometer for measuring the minimum of temperature during the twenty-four hours appears to be the one invented by Casella, Jun. Its peculiarity consists in having attached to the straight tube which indicates the temperature one of larger dimensions, terminating in a small chamber, by means of which contrivance the mercury, when it expands by heat, finds an easier passage into the chamber than into the tube itself, so that it remains in the tube at the same point to which it had receded during the cold of the preceding night, and thus indicates the lowest point to which it had sunk since the last observation.

But the accuracy of this method of determining the mean temperature of a place by ascertaining the diurnal maximum and minimum, depends upon the assumption, that the passage from the one to the other extreme goes on at a regular ratio throughout the twenty-four hours.

Thus it would not hold good if the diminution of temperature proceeded for a greater number of hours slowly, and afterwards went on at a more rapid rate for the remaining ones. Accordingly it has been found that this method is only quite exact for December, and that for the other months it requires a certain correction; no less than 1.9 being subtracted for July, 1.7 for August and May, 1.5 for April, 1.3 for September, 1.0 for October and March, and a small fraction of a degree for the remaining months.

Others therefore prefer the method of obtaining the mean temperature of a place by selecting for its registration some hour which has been found by previous observation to represent as nearly as possible the mean temperature of the whole day.

This varies a little for each place and for every month in the year; but if we take Quetelet as our authority, and regard the law he has laid down for Brussels as applicable to this and other adjacent countries, we may assign the hours from 8.12 to 9.36 in the morning, and from 6.40 to 8.6 in the evening, as the times most suitable for obtaining the mean temperature of the whole day. Accordingly the hours of nine in the morning and of nine in the evening are generally selected for observations on mean temperatures, but it must be recollected that for perfect precision certain corrections will be required, and probably observations taken at other hours should at the same time be registered[e].

Accordingly, Professor Dove of Berlin, who, as has been already remarked, is regarded as the highest authority on the subject of meteorology, has framed a series of most elaborate tables, which give the mean temperature of more than 500 places, respecting which sufficiently exact observations have been obtained. A selection from his reports have been made by myself, and by reference to the table thus constructed it will be seen, that in Europe most of the places quoted exceed the calculation made as to what might be called their normal temperature, or that due to their respective latitudes, whereas many in America and in Central Asia fall short of it.

Thus Petersburg, in latitude 59° 50', ought to have a mean temperature of about 30° 50', whereas observation assigns to it one of 39° 60', or 9° 10' in excess; Upsal, in the same latitude, one of 42° 54', or 11° 54' higher than that calculated; Copenhagen, in 55° 41', which should have a mean temperature of 35°, is found to have one of 46° 56', or 11° 56' higher; Edinburgh, standing nearly in the same latitude, a mean temperature of 47° 15', or 12° 15' above the mark; London, in lat. 51° 30', instead of 39° of Fahr., has one of 50° 83', or 11° 83' higher; Penzance, in 50° 7' north latitude, which would be set down at 41° of Fahr., enjoys a temperature as high as 49° 63'; and Gosport, in the same parallel, one of 51° 82', or 11° 32' above the calculated mean.

On the other hand, Yakoutzk in Siberia, in lat. 62° 1', has a mean temperature of only 13° 43', whereas its normal one would be 27° 0', or 13° 57' higher.

Irkutzk, in the same country, in the 52nd parallel, a mean temperature of 32° 62', or 6° 38' lower than it ought by calculation to possess.

[e] For these niceties, however, I shall refer to Drewe's practical work on Meteorology, or to Kämtz's German treatise, of which an abridgment has been published by Walker, translated into the English language.

Nertschinsk, in the 51st parallel, one no less than 15° too low, being quoted at 24° 17′ of Fahr.; Orenburg, in lat. 50° 45′, 6° too low, or 35°.

Now it is remarkable that these parts of Siberia lie near one of the two points determined by Hansteen at which there is no variation in the compass caused by the magnetic currents which circulate round the globe[f].

This curious coincidence between the extreme of cold and the *isoclinal magnetic lines*, holds good also on the opposite side of the globe, namely, in North America, although in a less striking manner, for the coldest spots in that hemisphere also are placed round and about the point which has been determined to be that of no variation.

Everywhere indeed in North America we find the temperature relatively lower than in Europe.

Thus Quebec, in lat. 46° 48′, possesses a mean temperature of only 41° 85′, which is a fraction of a degree lower than calculation would assign to it; and although in more southerly localities the observed exceeds the normal temperature somewhat, yet the excess is far less than we find to be the case in Europe.

Contrast, for instance, the temperature of places situated in corresponding latitudes near the western coasts of Europe and the eastern of America, — of Nantes, for instance, in France, and St. John's in Newfoundland, the former 54° 90′, the latter 38° 39′; or of Quebec and Poitiers, the first 41° 85′, the latter 53° 13′,—and it will be seen that the difference of position in the western or eastern sides of a continent has produced a difference of 16° 5′ of temperature in the former instance, and of 11° 28′ in the latter.

Yet even in these cases some slight exaltation seems to be due to the proximity of the sea, for in the interior of a vast continent, such as Asia, we have seen that the mean temperature is lower still, so as in some places even to fall as much short of the normal point, as on the coasts of the American continent it exceeds it.

[f] See in Berghaus' Physical Atlas, 4th Abth., Magnetismus, 5th No., or in Keith Johnson's Atlas, Erman's Map of the degrees of Declination from 1827 to 1831.

Nor is this all, for if we examine the mean winter temperature of these several places, it will be found that the discrepancy is still more striking.

Thus, whilst the winter temperature of Drontheim in the 63rd parallel is + 23° 29', that of Yakoutzk in the 62nd is — 36° 37', an enormous difference of 60°; whilst the winter temperature of Quebec is 14° 15', that of Rochelle in France, situated in the same latitude, is 51° 70'; and Penzance, 4° nearer the pole than either, enjoys a mean temperature of 44° 23', or one of 30° higher than that of Quebec.

Perhaps these conclusions may be rendered more palpable by exhibiting the following tabular view of the annual, the winter, and the summer mean temperatures of a few selected places on the surface of the globe, or of what meteorologists choose to denominate their *isocheimal* and *isotheral*, as well as their *isothermal* temperatures; the latter of which terms is confined to the expression of their mean heat during the entire year, whilst the two former terms indicate their mean winter and mean summer temperatures.

TABLE OF MEAN TEMPERATURES.

Locality.	N. Latitude.	Temperature. Annual. Real.	Normal.	Difference.	Winter.	Summer.
	DEG. MIN.	DEG. MIN.	DEG. MIN.	DEG. MIN.	DEG. MIN.	DEG. MIN.
EUROPE.						
Archangel .	64.32	33.53	25.00[g]	8.53	9.43	57.85
Petersburgh	59.50	39.60	30.50	9.10	18.66	61.68
Stockholm .	59.21	42.27	30.50	11.77	26.24	60.43
Upsal . .	59.00	42.54	31.00	11.54	23.76	59.17
Stromness .	58.59	46.54	32.00	14.54	39.35	54.42
Edinburgh .	55.58	47.15	35.00	12.15	38.45	57.17
Moscow . .	55.45	40.00	35.00	5.00	15.20	63.97
Copenhagen	55.41	46.56	35.00	11.56	31.31	62.70
Amsterdam .	52.23	49.86	39.00	10.86	35.63	64.39
Warsaw . .	52.13	44.15	39.00	5.15	25.20	64.60
London . .	51.30	50.83	39.00	11.83	39.50	62.93
Dresden . .	51.03	49.10	39.50	9.60	32.06	66.00
Gosport . .	50.47	51.82	40.50	11.32	40.97	62.74
Penzance .	50.07	49.63	41.00	8.65	44.23	60.91
Munich . .	48.80	48.38	44.00	4.38	32.50	63.65
Paris. . .	48.50	51.31	44.50	6.81	37.85	64.58
Vienna . .	48.13	51.03	45.00	6.03	31.95	69.40
Pesth . .	47.29	47.48	46.00	1.48	27.69	65.78
Montpellier	43.36	59.51	51.50	8.01	44.23	75.95
ASIA.						
Yakoutzk .	62.01	13.43	27.00	—13.57	—36.37	61.72
Irkutzk . .	52.19	32.62	39.00	— 6.38	0.90	61.50
Nertschinsk	51.18	24.17	39.50	—15.33	—16.83	61.10
Orenburg .	50.45	35.00	41.00	— 6.00	4.33	63.99
Astrachan .	46.21	50.05	47.00	3.00	19.17	75.94
N. AMERICA.						
Quebec . .	46.48	41.85	47.00	— 6.15	14.15	68.08
Montreal .	45.31	42.35	49.00	— 6.65	17.79	71.40
Rochester .	43.08	44.36	52.00	— 7.64	26.50	67.19
Toronto . .	43.40	44.81	52.00	— 7.19	25.43	64.63
Cambridge, (Boston)	42.25	48.61	52.00	— 3.39	27.34	70.15
New York .	40.00	51.58	55.50	— 3.92	30.12	70.93
Philadelphia	39.57	50.78	57.56	— 6.72	30.07	71.36
Cincinnati .	39.06	53.81	58.00	— 4.19	31.93	73.20
Washington	38.57	56.89	58.50	— 1.61	37.76	76.74
St. Louis .	38.36	55.16	58.56	— 3.34	32.57	75.27

[g] These figures must only be regarded as approximations to the truth.

Climates, then, may be divided into equable and excessive, according to the degree in which the mean temperature of the summer and winter differs from that of the entire year; and with reference to the growth of vegetables, far more importance must be attached to the heat prevailing during summer than to the mean temperature of the climate collectively taken.

Thus, as will be more fully explained in a subsequent lecture, even in Russia and Siberia fine crops of wheat and other kinds of corn are obtained, because the summer temperature rises to the requisite point for ripening the seed, whilst in the north of Scotland, the Orkneys, and the Faroe Islands, although the mean temperature of the year is higher, these crops do not succeed.

It may be doubted, however, whether even the mean temperature of the summer season affords a sufficient clue to all the variations in the character of the vegetation which are attributable to heat. A plant is not like a spring, which is pushed forwards a certain number of degrees by the application of a definite force, and when that pressure is removed, returns again to its original position; for when the stimulus of heat is applied to it, its organs undergo a degree of development, which they retain even although the temperature should afterwards be reduced. Hence it is necessary to note the extremes of temperature to which a country is liable, as well as the mean of its summer and winter climate.

Should it happen, for instance, that the cold in a low or sub-tropical latitude ever approaches even for a single night to the zero of Fahrenheit, certain trees, such as the Orange, would infallibly perish; and hence they can never be indigenous in countries subject to such contingencies.

The general climate of the British Isles is so exceptionally mild, that we have introduced the plants of warmer regions generally into cultivation, and begun to consider them as in a manner naturalized; but that they are not so, and could never have established themselves in the soil without the aid of man, became evident from the effects of the rigorous winter of 1860-61,—one of those seasons of unusual severity which, however, are sure to recur within a certain cycle of years, and to entail the destruction of all those denizens of a more temperate climate, which, rashly presuming upon the mildness of many preceding winters, had begun to regard this as their home.

In the season to which I allude the thermometer fell in Cambridgeshire to — 15° below zero; in several of the midland counties to — 12°; in Oxfordshire and Gloucestershire to — 2°; and even at Dawlish to the unusual point of 8° of Fahrenheit.

No winter at all approaching this in point of severity occurred in England since that of 1837-8, when the thermometer at Walton, near Claremont, is quoted at — 14°; at Bicton, in Devonshire, at 18°; and at Binstead, in the Isle of Wight, at 15° of Fahr.

In that year the temperature in Cambridgeshire is stated to have been — 3°; at Chiswick, — 4°; in Norfolk, — 3°; and in Surrey as low as — 14°.

These occasional invasions of extreme cold tend, of course, to curtail the number of trees and shrubs which can be introduced from more southern latitudes, even though capable of enduring without injury the ordinary severity of an English winter.

But the destruction of tender evergreens and other plants on the late occasion seems to have been out of proportion to the difference between the cold in this and in former severe seasons. In both cases, indeed, the Laurels of all kinds—*Arbutus unedo, Photinias, Edwardsias; Pinus longifolia, insignis, halepensis,* and others, were killed to the ground; but in the instance alluded to we had to deplore in many of our midland counties the probable loss of the *Deodaras,* which we had flattered ourselves would have been a permanent accession to British timber-trees, and of several which had lived for many years past in the Oxford Botanic Garden, such as the Judas-tree, the *Arbutus Andrachne,* &c.

It was the peculiar severity of a few days in this winter, rather than the greater coldness of the year itself as compared to others, that caused this destruction of the cultivated plants in our gardens and pleasure-grounds, for the mean temperature of 1860-1 appears to have been only 3° 13' below the average; although if we reckoned that part of the winter came into the next year, the rigour of the season would be set down as somewhat greater.

Unusual, however, as the cold in 1860 and 1837 was for the British Isles, it never reached the point ordinarily arrived at in much lower latitudes of America, for 20° below zero is not unfrequent in Pennsylvania and New Jersey, places where the thermometer rises in summer to 100° and even 110° of Fahrenheit in the shade.

Hence we can understand the absence of our common evergreens, and even of the comparatively hardy Ivy, which strikes us at first with so much astonishment in the midland states of North America; in latitudes, that is, where the Maize flourishes, and where an almost tropical heat is felt during a portion of the year.

On the other hand, a hardy plant like the Vine seldom succeeds in England; but whenever it meets with a certain number of days of sunshine of an intensity sufficient to mature its fruit, the vintage will be good, not only if the winter be cooler, but even though the mean temperature of the year be lower than is the case in England. Thus the grapes at Astrachan are said to be delicious, and yet the mean winter temperature is 19°, and that of the year 50°, which is 1° lower than that of London.

Now the differences in climate, which have just been pointed out, may be referred to two sets of causes, the first of which are general, the latter local.

The general causes which affect temperature are, the relative distribution of sea and land; the proximity to an extensive tract of continent, or to a wide expanse of ocean; the shelter afforded by forests, or hills of moderate elevation; and the chilling influence of a chain of mountains sufficiently lofty to retain the winter snow during the greater part of the year, or to give rise to glaciers which may invade the neighbouring valleys.

In order to understand how these circumstances affect climate, it will be necessary to consider the different manner in which the solar rays operate upon the surface of land and of water, first in tropical, and secondly in arctic regions.

Now it is evident, that wherever the sun strikes so directly upon the earth as to produce a powerful effect, the surface soil will get hotter and hotter as the day advances, owing to the absorption of the solar rays, and the slow transmission downwards of the heat thus generated, by materials so deficient in conducting power as those of which the crust of the earth is for the most part composed.

So feeble, indeed, is the conducting power of the ground upon which we tread, that in Siberia, in places where the heat in summer rises sometimes to 90° and upwards of Fahrenheit, the soil remains frozen throughout the year to a depth, according to Middendorff, of more than 380 ft., for at Yakutzk, in latitude 62° 2′, the temperature, which at 50 ft. from the surface was about 17° of Fahrenheit, had only risen to 26° 6′ at the above depth [g].

And yet even here there can be no doubt, that at a still lower point a high temperature is attained; and that this augmentation goes on in all parts of the globe at a certain definite rate down to the lowest point to which man has ever penetrated,—arising, as is supposed, from the internal heat of the globe.

Accordingly, if we set aside the cold produced by the evaporation of moisture from the surface, as well as the heat abstracted from the ground by the stratum of air which immediately touches it, and which, as it becomes heated, expands, and rising gives place to another and a colder portion; we shall see reason to attribute the cooling of the solid portion of the crust of the globe mainly to radiation from its surface upwards towards the sky.

Now as the interchange of heat between two bodies by radiation depends upon the relative temperature which they respectively possess, the earth, by the rays transmitted from the sun during the day, must be continually gaining an accession of heat, which would be far from being counterbalanced by the opposite effect of its own radiation into space.

[g] Humb. Cosm., vol. iv. p. 44.

Hence from sun-rise till two or three hours after mid-day, the earth goes on gradually increasing in temperature, the augmentation being greatest, where the surface consists of materials calculated from their colour and texture to absorb heat, and where it is deficient in moisture, which by its evaporation would have a tendency to diminish it.

In regions where the sun's rays are powerful, and the atmosphere sufficiently transparent to allow of their ready transmission, the propinquity of hills of moderate elevation may even augment the temperature of the lower levels, by radiating heat downwards upon them, if their own surface be black and of an absorbent texture, or by reflecting it, if they be of a lighter colour and of a more glistening character.

These effects will, however, be counteracted by the radiation into space of heat from the ground, which continues uncompensated by that from the sun, after darkness has set in.

If, indeed, the sky be overcast, there will be some return of caloric by counter-radiation from the clouds; but when the atmosphere is clear and transparent, the heat transmitted from the surface into space during the night may produce a great reduction of temperature, sufficient indeed in some cases to cause water even in tropical latitudes to congeal. Indeed, it is from this cause that ice is procurable in Bengal, by exposing water in shallow earthen pans to the night air.

But to this part of the subject we shall return in the next lecture.

Far different is the influence of solar heat upon the surface of water in corresponding latitudes. Although the radiating properties of the fluid are inferior to those of the solid portion of the globe, the power of distributing heat belonging to the former is nevertheless greater, for whilst a part of the rays which impinge upon the surface of water is reflected, another portion is transmitted downwards through the body of the fluid, and becomes gradually absorbed in its passage through it.

Now the heat absorbed by a fluid increases the rate of its evaporation, and the latter process reduces the sensible, by increasing the amount of latent heat which the body requires.

Hence every accession of temperature brings with it its own remedy, by adding to the amount of aqueous vapour disengaged from the surface of the liquid; and, accordingly, during the day the temperature of the water never rises much beyond the mean of the climate at the particular season.

But at night the case is reversed, not only because radiation takes place more slowly from the surface of water than of land, but also because the former, when cooled, becoming heavier and sinking, gives place to another portion from below, so that the temperature at the surface continues almost unaltered.

Hence it is that islands, and other maritime tracts which partake of the temperature of the contiguous seas, enjoy an equable climate, whilst extensive continents at a distance from large bodies of water possess an excessive one.

And hence, too, it happens, that in warm latitudes, there is always a cool breeze setting in from the sea during the day, and the same from the land during the night, for the latter being hottest by day, produces an upward current of air, which causes a rushing in of cooler air from the sea to supply its place; and for the same reason becoming coldest after sunset, brings about just the reverse effect during the night.

Let us now consider the mode in which the solar rays affect the surface of the land and of the sea in higher latitudes.

In summer, of course, the same general difference between the two with respect to the absorption and emission of heat will prevail, but wherever a chain of mountains of a certain elevation exists, the climate will be rendered more rigorous in the valleys and plains below, owing to the interchange of caloric between the latter and the high ground near, as well as to the conversion of the sensible heat received by radiation into a latent form, which is caused by the melting of the snow and ice covering the slopes and summits of the hills.

Thus it is only on extensive plains, at a distance from snow-capped mountains, that a high range of temperature can maintain itself even during the day in a northern latitude. On the other hand, over such level tracts as those of Russia and Siberia, great heats prevail in summer even in comparatively northern regions.

Moscow, for instance, in north latitude 55° 48', where the mean temperature of the coldest month is only 13° of Fahr., enjoys in July a heat of 66° 4'; and at Yakoutzk, a Siberian town, in latitude 62°, at which the mean temperature of January is 36° 37' below zero, the thermometer in July rises to nearly 69° of Fahrenheit.

In islands, on the contrary, situated in northern parallels, the radiation of heat in summer exerts a much feebler influence; so that Stromness, in lat. 58° 57', has a summer temperature of only 54°; and Unst, the most northern of the Shetland Islands, in lat. 60°, one only of 52°.

Hence whilst at Christiana, in lat. 59° 55', fine timber abounds, and crops of wheat and other grain ripen, inasmuch as the mean summer temperature rises nearly to 60°, none but the hardiest kind of barley will grow in the Hebrides, between the parallels of 56° and 58°; and the trees are there reduced to a few of the robuster species, which are both stunted and uncommon.

In winter, however, the case is reversed. The extreme cold of extensive continents, such as Russia and Siberia, is caused in part by the radiation of heat from their surface during the long nights of these northern latitudes, and still more through their participation in the climate of regions more northerly than themselves, owing to the winds which commonly come from that quarter in the winter season, and which, bringing with them the temperature of the Arctic circle, first condense the moisture into snow, and afterwards impart to the countries they pass over the dry and cutting cold which characterizes them.

But on the sea the circumstances are different, owing to a property peculiar to water, which would seem specially designed as a provision for mitigating the intensity of cold.

This is its arriving at its greatest density, not at the point at which it freezes, but 8° above, so that whilst it goes on progressively contracting in volume down to 40°, it afterwards again expands, until it falls to the temperature of 32°.

Let us consider how this circumstance affects the cooling of large bodies of water.

Supposing the liquid at the commencement to be at 50° or 60° of Fahr., the access of cold from the north will gradually abstract its heat, until by the sinking of the heavier, and the rise of the lighter strata of the fluid, all the upper portions are brought down to the standard of 40°.

But when this is effected, no further diminution of temperature can take place, until the whole body of the liquid has sunk to the same level, because the instant it reaches that point, its greater density causes it to gravitate below the lighter and warmer water which lies beneath it.

Thus supposing the entire ocean in that part to be reduced to 41°, it could not be brought below 40° on the surface by any degree of cold which might exist in the atmosphere above, until the whole body of water had sunk to that level, because the instant it had reached it, it would gravitate to the bottom, and be replaced by the warmer water beneath.

Hence owing to this constant circulation of the lighter and heavier portions of the water, the whole must attain the temperature of 40° before any ice is formed upon its surface, and accordingly it is hardly possible, that a deep lake should be frozen over even by the longest and most intense frost that can occur.

The effect of this constant circulation throughout all portions of a body of water in mitigating the severity of insular climates is sufficiently apparent.

The ocean may in fact be regarded as a store-house of heat, which it dispenses to the air passing over its surface, thus rendering it impossible that the latter should ever attain the same extreme degree of cold which it acquires on a continent.

Hence the equable character of the climate in insular situations, which has been pointed out as prevailing during the summer, holds good also, for the reasons just given, in the winter likewise.

It has even been conjectured, that if a large expanse of water existed at either pole, it would continue from this cause uncongealed, for no sooner had the surface arrived at 40°, than it would gravitate downwards, and be carried by an under current towards the equator, its place being supplied by the lighter and warmer water of the tropics, which would be moving northwards.

This speculation became the more interesting, from the discovery, which Dr. Kane the American Arctic navigator, professed to have made, of an open polar sea, which arrested the progress of his exploring party to the north, and which was surveyed from an elevated point of land called by them Cape Constitution, which he pronounces to be the highest northern land, not only of America, but of the globe. As they approached the coasts of this polar ocean, the ice upon which they had travelled became rotten, and the snow wet and pulpy. They found themselves on the shores of a channel so open, that a frigate, or a fleet of frigates, might have sailed up it. In every direction, so far as their eye extended, the waters appeared unencumbered with ice[h].

Without vouching for the truth of this statement, which has since been disputed, it is possible, that if no tract of land did exist between 80° of north latitude and the pole, the temperature would be comparatively mild, and the sea as navigable as in lower latitudes.

Even betwixt continents, indeed, a difference of climate exists, which may be best referred to the same circumstances.

Supposing, for instance, one, like America, to be connected with the polar regions by a continuous belt of land, it would possess a more rigorous climate than another like Europe, which is separated from them by a wide expanse of water. This is one of the causes of the vast difference in this respect between Great Britain and Labrador, countries placed in nearly corresponding latitudes.

How far a different distribution of sea and land from the

[h] Vol. i. p. 302.

present might affect the general temperature of the earth's crust has been a favourite speculation with geologists, who in every step of their inquiries are met with the startling fact, that the inhabitants of our former seas, and the tenants of whatever dry land might have then existed, whether belonging to the animal or to the vegetable creation, represent more nearly the fauna and flora of tropical, or at least sub-tropical, regions, than those of the more northerly latitudes in which they are so frequently found.

It is impossible, I think, to resist this conclusion; for although it may be true, that Tree-ferns are met with as low as New Zealand[1]; that one species of Palm grows in Tasmania, and another at a height of 8,000 feet on the Himalayas; and although the larger quadrupeds of the tropics may exist for a time in the colder regions of the north, yet these must be regarded as stragglers from their normal position, not as representatives of the class of plants and animals appropriate to the country and latitude.

And the broad fact which meets us, whenever we examine into the records of the creation, as displayed in the organic remains, animal or vegetable, that have been preserved, is, that the temperature during the whole of the secondary, and most of the tertiary epochs, was sufficiently exalted, at least as far north as latitude 55° or 56°, to admit of the growth of Tree-ferns, Cycadeæ, Araucariæ, gigantic Lycopodiaceæ, and in a few instances even of Palms; whilst Reptiles of the kinds that now are confined to the warmer regions of the globe existed as low down as the chalk, and Coral reefs continued to be formed down to the commencement of that great accession of cold, which gave rise to what geologists have called the Glacial Period.

And this is rendered more striking, when we reflect that the portion of the globe we inhabit is even now enjoying a temperature perhaps belonging normally to a latitude 10° lower, so that it is not difficult to understand that changes

---

[1] Hooker states that the most southern latitude in which Tree-ferns have been found is in the north of New Zealand, south latitude 44°. Phillips, however, mentions *Aspidium arboreum* in 52° 5', as well as the *Alsophila australis* in Van Diemen's Land, and the *Cyathea Dicksoni* in New Zealand.

in the distribution of sea and land might have brought about the excessive cold which appears to have prevailed just before man came into existence.

Sir Charles Lyell has suggested, that the most favourable condition for diffusing throughout the globe a genial temperature, would be that, in which all the principal tracts of land were collected within the tropics, and all the water in more northern and southern latitudes, as the soil heated by the solar rays would communicate its warmth to the surrounding waters, and thus produce a northerly and southerly current, which would modify very materially the climate of the Arctic regions, whilst the latter, consisting of water, would, for the reasons above stated, undergo a less degree of refrigeration than at present.

This theory is developed by the author with his usual ability, but it may be questioned, whether it be not beset with unnecessary difficulties, owing to his adherence to his favourite principle of *uniformitarianism,* or, in other words, his reluctance to admit, that any progression can be traced in the order of physical events from the earliest to the most modern condition of our planet.

And yet the doctrine of Darwin, which he now espouses, would seem more in harmony with such a notion, if even it does not imply it, than that of the reversion of the earth's surface to its *status quo,* after the fulfilment of a certain cycle of revolutions; for if the globe had been equally fitted for the abode of man and the higher Mammalia from the earliest period known to us, as it is at the present time, one does not understand why such classes of animals should have been absent, and if they did exist, what then becomes of the theory which supposes the gradual evolution of more complex from more simple organisms, of man from the monad?

Sir Charles Lyell has been accused, most unjustly, of maintaining that the world never had a beginning, but this erroneous impression of his meaning has arisen, from his contending, that geology has never yet mounted far enough into the records of creation to be able to find traces of its dawn, or, in other words, that from the earliest to the latest of the deposits which compose the crust we recognise only a series of alternating movements, of elevations of land in one region, and of depressions in another; so that in the course of an indefinite number of ages each portion of the globe will have been subjected at one time or another to the same internal commotions, occasioning the same amount of external change.

Now if this view be correct, one does not see, why periods of intense cold should not have been intercalated at many different epochs between those of high temperature, which the organic remains preserved in the rocks appear to indicate, and why evidences of glacial action anterior to the post-pliocene epoch should not have been traced repeatedly.

One such event, indeed, has been pointed out to us by Professor Ramsey, who contends, that towards the close of the palæozoic period a glacial sea was spread around several of the islands which occupied the place of what now constitutes our British and Scottish mountains, and that the remarkable conglomerate rock found in the Malvern and Abberley Hills, now pronounced to be of permean origin, is due to the floating of icebergs along the sea which washed the Langwyd, Abberley, and Malvern Hills.

Granting, however, the correctness of this deduction, it still remains to be explained why, if the *sources* of heat in former periods did not possess more intensity than at present, the local circumstances should have been, in so much the greater number of instances, more favourable to their operation than is the case under existing conditions.

It seems indeed contrary to all probability, that the very arrangement of sea and land, most favourable to the production of warmth, should have existed almost universally, till the commencement of the glacial epoch, which represents nearly the most modern date in our geological calendar.

But what would be the temperature of the entire globe, if it had been uniformly covered with sea, so that the flow of its warm currents proceeded continually from the equator towards the poles without let or hindrance from the interposition of continents, and that at a time when, owing to the absence of land, no such radiation of heat into space occurred, as takes place from continents at present?

Without pretending to calculate what the precise temperature in each parallel would be under such circumstances, as we have probably not sufficient data for determining this with exactness, it may at least be affirmed, that the climate everywhere would be much more uniform, and that even in the polar regions it would nowhere at the surface sink to the freezing point of water; nor would the case be materially altered, even when large islands started up in the midst of the abyss, provided the latter were not numerous enough to oppose a barrier to the circulation of the aqueous currents, and also presented no lofty chains of mountains to serve as centres of refrigerating influence.

It may be remarked, that the plants that have principally left their traces in the coal formation, namely the Ferns, Lycopodiaceæ, and the like, are such as abound principally on islands, and delight in an equable and humid, rather than in a scorching and sultry atmosphere. Even the Tree-ferns of the tropics thrive principally in the midst of deep forests, protected at once from drought and from excessive heat.

If therefore we suppose the globe to have been at that time covered with water, with only some islands occasionally protruding above it, it is conceivable, that the temperature even in these latitudes might have been exalted enough for the plants that have left their vestiges in the coal.

Now this gradual emergence of the land above the waters is rendered probable, not only from considering the nature of the plants and animals which came successively into being, but also from the character of the rocks themselves which predominated at each successive period.

With regard to the former, it may be remarked, that at the earliest epoch of which we find any traces, namely the palæozoic, fish and marine productions alone existed; at one somewhat later, during the carboniferous, reptiles were the only animals that peopled the land; whilst mammals of a low order first appeared in the oolite, and went on gradually increasing in number and in variety in proportion as the most recent period was approached.

And with reference to the rocks themselves, it seems unquestionable, that all those of igneous formation which have been observed intruding themselves into the strata, appear to be submarine, for although it may be said, that all traces, of craters, of cones of scoriæ, and even of streams of lava, might have been obliterated by the series of catastrophes that had since occurred, yet vestiges of the description of rocks which had constituted them would nevertheless present themselves, if volcanos had at that period broken out frequently upon dry land.

Their *débris*, even if washed into the contiguous seas, and covered over with neptunian deposits, would have retained their vitreous and porous character—they would have partaken often of the structure belonging to obsidian and pumice, which indicates sudden cooling; whereas the igneous rocks we observe in the older formations, if cellular, have their interstices filled by crystaline matter, and possess when compact that lithoid aspect which is so generally absent from lavas erupted in the open air[k].

Thus, whether we consider the nature of the rocks themselves, or that of the organic beings that existed upon their face, we are alike led to the conclusion, that the globe was at first covered with water,—that, next, only a few low islands gradually emerged from the abyss, suitable for the abode of reptiles and other of the lower animals,—that these gradually increased in number, and became more elevated in position—but that until a later period no such extensive continents or such lofty chains of mountains existed, as would be sufficient to bring about that degree of cold which belongs to the higher latitudes of the globe at present.

This low temperature, since the introduction of man upon our planet, has been mitigated by causes already pointed out in this lecture, but it is quite conceivable that, supposing a different disposition of land and sea to have existed antecedently, a climate might have prevailed in these regions as severe as that of Siberia, and therefore corresponding to what we find traces of during the so-called Glacial Period.

[k] See my "Description of Active and Extinct Volcanos," 2nd Edit., 1848; especially chap. xl. p. 669, &c.

LECTURE II.

Local influences affecting temperature—the Gulf-stream increasing the warmth of the West of Europe—the Antarctic current increasing the cold of the Southern Hemisphere. Cause of the greater coldness of mountainous regions. Line of perpetual congelation. Temperature of the soil differs from that of the atmosphere—its influence on vegetation. Geothermal culture—instance of its advantages. Influence of solar light as distinct from solar heat. Instruments for measuring it. Humidity of a climate. Instruments for estimating it. Productive of three classes of phenomena: first, Fogs, mists, and clouds—nature of vesicular vapour; second, Rain and Snow—causes producing them—why rain does not fall in certain countries: third, Dew—theory of its formation. Winds. General laws affecting the aerial currents—arising from the difference in the temperature of different zones—coupled with the effect of the earth's rotation. Hence the Trade Winds. Why calms exist near the equator. Monsoons, typhoons, and hurricanes—their vorticose and progressive movement. Winds in temperate regions more variable and less violent, but apparently subject to the same laws—Admiral Fitzroy's method of forecasting their arrival. Influence of winds on vegetables. The Simoom. Pressure of the atmosphere varies in proportion to the elevation—hence the latter may be estimated by the barometer. The pocket aneroid, the most convenient and portable instrument for estimating heights in travelling. Pressure of the atmosphere dependent on height. Indications afforded by the barometer. Electrical state of the atmosphere during rains—hail dependent on the same cause—its effects on plants and animals. Ozone—how its presence is detected—its probable nature—its existence in certain states of the atmosphere—and the inferences to be deduced from its presence.

In my former Lecture I considered the general causes which influence climate, namely, the proximity to the sea, the existence of high mountains, the radiation from a large surface of sandy desert, and the like.

I shall now point out to you certain causes of a more local kind which affect the temperature of particular parts of the globe, and shall begin with one which seems to be chiefly instrumental in bringing about that mild and equable character of the seasons for which this spot, in common with most parts of the western coasts of Great Britain and Ireland, is celebrated.

I allude to the Gulf-stream, a branch of the great equatorial current, which, as will be explained when speaking of the trade winds, is constantly circulating round the globe from north-east to south-west above the equator, and from south-east to north-west below it.

This tendency of the waters of the ocean to the west, produced, as is supposed, by the influence of the trade winds, is interrupted owing to the barrier opposed by the American continent, and hence the north-easterly current is deflected by the coasts of the Gulf of Mexico, and compelled to flow through the strait intervening between the southern part of Florida and the West India Islands, from whence it emerges into the open sea.

Here it is first known as the Gulf-stream, and under this name flows across the Atlantic in a diagonal direction, so as to impinge upon the great bank of Newfoundland, where it is recognised by the higher temperature it imparts to its waters in comparison with those of the surrounding ocean.

It then travels eastward to the coasts of Europe, and produces a very sensible effect upon the climate of Great Britain, and even of Norway, contributing mainly to bring about that remarkable difference which has been already noticed in my last Lecture, as existing between the temperature of regions so nearly corresponding in point of latitude as Devonshire and Labrador.

Its influence is felt even more powerfully in Cornwall, especially at Falmouth and Penzance, admitting in both places the growth of plants of too tender a nature to stand the winter even in the most favoured spots in Devonshire, and rendering the latter locality famous for the early growth of vegetables intended to supply the London markets.

It is to the Scilly Islands, however, that we must look for what is due to the unmixed influence of the Gulf-stream, lying, as they do, so detached from the mainland, as to be but little affected by its temperature.

Trees, indeed, are scarcely found on these islands, owing to the boisterous winds that prevail; but in spots protected from their violence, shrubs and herbaceous plants grow with the utmost luxuriance, and evince by their nature the extraordinary mildness of the climate [a].

The Gulf-stream when issuing from the straits of Florida has an average velocity of four miles an hour, and a temperature of 86° Fahr. In the parallel of 42° or 43° it ranges still between 76° to 79°, and there by its genial warmth favours the rapid growth of that species of sea-weed called the Sargasso, or Gulf-weed, which covers vast tracts of sea in those latitudes.

Even on our own coasts, though its temperature is much reduced, it produces those warm vapours which moderate the rigour of our winters, and sometimes indicates its presence by conveying to us tropical plants, as well as multitudes of sharks, which is reported to have been the case during the last season.

I will next point out to you an oceanic current which seems to operate upon other parts of the globe in quite the reverse manner to the Gulf-stream.

Callao, on the western coast of South America, is situated in the 12th parallel of south latitude, whilst Rio Janeiro, on the eastern coast, is in the 23rd. Yet the former has a mean temperature of only 68° 9′ of Fahr. (20 cent.), whilst the latter has one of 73° 76′, (23° 26′). In like manner the Falkland Islands, in south lat. 52°, have a mean temperature of 46° 83′, (8° 24′ cent.); and Port Famine on the coast of Magellan, in south lat. 53°, one of 41° 5′, (5° 3′); whilst the Faroe Islands, in north lat. 62°, have a temperature of 45° Fahr., (7° 1′); and Dublin, in north lat. 53°, one of 49° 3′, (9° 6′).

Now the lower temperature of the shores of the southern hemisphere, as compared with the northern, seems to be owing to the influx of the waters of the Antarctic Ocean, unmitigated by any such genial current as the Gulf-stream. A cold stream flowing from the Polar seas reaches even the latitude in which Callao is situated, before it takes a westerly direction, and acquires the temperature of the general body of the Pacific.

[a] See Appendix.

Obtaining, however, by degrees a heat of 80° or 82° of Fahr., it flows into the Indian Ocean, and a portion reaching the eastern coast of Japan, imparts to those islands a more genial climate than belongs to their position on the globe.

Thus far I have confined myself to the question, how far the temperature of different localities is affected by their position, without reference to their respective height.

Let us now briefly inquire into the differences in that respect which are produced by a high or low elevation.

We are all familiar with the fact, that the air of mountains is cooler and fresher than that of the plains below, and that every addition to the height above the level of the sea brings with it a corresponding reduction in the temperature of the place itself. Hence in all latitudes, even indeed at the equator, there is a certain point at which perpetual snow prevails, the sun's rays being never powerful enough to maintain the highly rarefied air which exists at such an elevation above the point at which water congeals. Indeed, if other causes did not interfere, we might estimate the relative height of two places by simply noticing the difference of temperature subsisting between the two.

The late balloon ascents of Mr. Glaisher, of which he gave us an account the other day, seem, however, to shew that we have still much to learn on this subject; for it would appear from his statement, that the law of decrease of temperature does not go on with any regularity in the upper regions of the atmosphere; the fall in the thermometer for the first 5,000 ft. being no less that 20° of Fahr.; sinking to 10° between 5,000 and 10,000 ft.; from thence to 15,000 ft. being only 7°; and lastly, from 15,000 to 25,000 ft. only $2\frac{1}{2}$°; after which the temperature sunk 9° between 25,000 and 30,000 ft., the highest point attained by this aeronaut.

Further observations, therefore, are necessary to reconcile these anomalies with the received opinions on this subject. They arise, no doubt, from the intermixture in the higher regions of the atmosphere of aerial currents of different temperatures, by which we may account for the fall of hail in summer, and for its occurrence even in more southern latitudes where frost is altogether unknown.

I know of no other assignable cause for the greater coldness of rarefied air, than the increased capacity for heat which the latter acquires in proportion to its expansion.

Upon the atmosphere, indeed, the solar rays will fail to produce their full effect, because they are only in part absorbed in their passage through it to the earth; but although this circumstance may account for the cold experienced in mounting into the higher regions of the atmosphere, it leaves it still for us to inquire, why an equal amount of solar light does not warm the surface of an elevated table-land, in the same degree as it will do a tract situated near the sea's level, such as the great plains of Egypt or Arabia. Be that, however, as it may, it cannot be denied, that the climate of a country, so far as regards temperature, greatly depends upon its physical elevation; and that although hills of moderate height in a warm latitude may even enhance the temperature of the adjacent plains, both by radiating heat down upon them during the day, and by obstructing its free passage upwards during the night, yet the proximity of mountains, lofty enough to be covered with snow during the summer months, will tend very much to depress it; just as a ball of ice placed in the focus of a metallic reflector causes a thermometer, situated in the focus of another mirror opposite to it, to sink rapidly.

But another element in the consideration of climate, as concerns the vegetable kingdom, is the temperature of the soil.

This is by no means uniformly the same as that of the external air, for we find the earth to be heated, at least for short periods of time, to a much higher point than the atmosphere above, so that at the Cape of Good Hope Sir John Herschel observed, that on the same day on which his thermometer in the shade ranged from 92° to 98°, the soil of his garden caused it to mount up to 150° and 159°, and even in shaded spots as high as 119°. Even at 5 P.M. the temperature of the soil, similarly circumstanced, was as high as 102°.

In other countries temperatures as much exceeding that of the air as these noticed are recorded, and the following are a few of the statements given, with respect to the temperature observed immediately beneath the earth's surface, in different parts of the globe:—

Country.	Temperature.	Authority.
Tropics, often	126—134°.	Humboldt.
Egypt	133—144°.	Edwards and Colin.
Oronoco, (air being 84° 5')	in white sand 140°.	Humboldt.
Chili	113—118°, among dry grass.	Boussingault.
Cape of Good Hope	150°, under the soil of a bulb garden.	Herschel.
Bermuda	142°, thermometer barely covered in earth.	Emmet.
China	water of the fields 113°, adjacent sand much higher, blackened sides of the boat at mid-day 142—150°.	Meyer.
France	118—122°, and in one instance 127°.	Arago.

The above, however, I presume, applies to places where the intensity of solar radiation is very intense; for whilst in India the difference of temperature between the earth and the air is always in favour of the former to the extent of 8° in summer, and about 3° in winter, in England it would appear that at Chiswick it did not exceed 2°, the earth being so much warmer than the air in autumn and winter, and nearly as much colder in May, June, and July; whilst in March and April the two nearly corresponded. And in accordance with the above results, the following table shews the mean monthly temperature at Chiswick, as indicated by a thermometer the bulb of which was plunged 24 ft. below the soil during the year 1857.

Assuming the mean temperature in London to be 50° 83',

the difference between the warmth of the ground and of the air in each month is represented by the signs — or + prefixed to the figures:

TEMPERATURE OF THE SOIL AT LONDON AS COMPARED TO THAT OF THE AIR, THE LATTER ASSUMED TO BE 50.83.

1857.	DEG. MIN.	DEG. MIN.
January	51.05	+ 0.22
February	50.26	— 0.57
March	49.37	— 1.46
April	48.65	— 2.18
May	48.24	— 2.59
June	48.23	— 2.60
July	48.72	— 2.11
August	49.64	— 1.19
September	50.68	— 0.15
October	51.69	+ 0.86
November	52.29	+ 1.46
December	52.31	+ 1.48

The importance of this to vegetation may be estimated by the following considerations.

It is known that every plant requires a certain amount of heat, varying in the case of each species, for the renewal of its growth at the commencement of the season.

Now when this degree of heat has spurred into activity those parts that are above ground, and caused them to elaborate the sap, it is necessary that the subterranean portions should at the same time be excited by the heat of the ground to absorb the materials which are to supply the plant with nourishment. Unless the latter function is provided for, the aerial portions of the plant will languish from want of food to assimilate. Indeed, it is even advisable that the roots should take the start of the leaves, in order to have in readiness a store of food for the latter to draw upon [b].

[b] That the roots are not merely passive agents, I have shewn in my paper in the Journal of the Chemical Society, vol. xiv.; and their importance in elaborating the materials for the development of the aerial portions of the plant has been insisted upon in Liebig's late work, entitled "The Natural Laws of Husbandry."

The practical importance of maintaining the soil at a higher temperature than the air, has been lately pointed out by a French naturalist, M. Naudin, who gives instances, in which the mere passage of spring water from a neighbouring rock through the bottom of a conservatory communicated heat enough to maintain its temperature above the freezing point, without the assistance of any artificial heat.

The object may, however, be effected more completely by means of flues heated at certain periods, and by affording protection to the plants in winter by matting, tarpaulin, &c., which latter will also contribute to preserve the stem and branches from the effects of the winter's frosts.

In the "Gardener's Chronicle" for Feb. 16, 1861, Lindley corroborates these statements of M. Naudin. "The mere protection," he says, "afforded by a glass frame will tend to keep up this bottom heat, for although it be true that glass radiates heat more quickly than textile fabrics, and consequently will tend in this way to lower the temperature of the ground, yet if the soil be merely excavated to the depth of six or eight feet, and glass be placed over it on a level with the surface of the ground, a thermometer suspended below the glass would not fall below freezing during the frosts of ordinary winters, and probably not even in one as intense as that we had in 1860. The heat which maintains a comparatively high temperature below the glass, can in this case only be derived from the natural warmth of the earth."

Indeed, it has been found, that many half-hardy plants might be kept alive by merely passing flues through the soil in an open border, so as to impart to their roots an artificial temperature, even though the plants themselves were unprotected by any glass at all. This method, which goes by the name of Geothermal culture, although suggested by Dr. Lindley in his work on Horticulture as long ago as 1855, does not appear to have been as yet put into practice to the extent which it deserves.

It is on this principle that Cape Bulbs thrive so much better in their own native soil than in stoves, for though it may be easy to raise the temperature of the air to any given point, it is not equally practicable to maintain the soil at the high degree of heat, which it attains, as we have seen, in the places where they are indigenous.

A remarkable instance of successful geothermal culture has been afforded by a Brazilian climber called the Bougainvillea speciosa, which was first made to flower freely in England by Mr. Keene's gardener at Swyncombe House, near Henley, in Oxfordshire, owing to his bringing the roots of the plant into close proximity with the flue of the stove, and thus obtaining a bottom heat almost sufficient to bake the soil in which they were placed.

This plant, from its numerous and graceful festoons of flowers, or rather bracteæ, like those of the female hop, but of a rich colour, in which crimson, violet, and purple are exquisitely blended, presented a most gorgeous spectacle, and excited great admiration in Oxford at a flower-show in 1861, where it was for the first time exhibited.

It has flowered in the same conservatory for three successive years, its blossoms first making their appearance in January, and continuing in the greatest exuberance till May or June[c].

But two countries may be very differently circumstanced as to climate, even though enjoying the same temperature, according to the amount of solar light which they respectively receive.

It is even conceivable, that the country least favoured in the former respect may be the most sultry, because the clouds, when they intercept the solar beams, contribute to the warmth received by the earth, by reflecting back the heat which the latter would lose by radiation.

Thus an Alpine valley in the summer season may be subjected to a heat more oppressive than a level plain contiguous, although it experience a less amount of light; and possibly the different influence it exerts upon plants and animals may be dependent upon this circumstance.

It is therefore important to have instruments for mea-

[c] Another species, B. glabra, also a beautiful plant, does not require bottom heat, and may be made to flower probably in any greenhouse.

suring the force of solar radiation, as distinct from the mere temperature of the place as observed in the shade, and for this purpose we employ a thermometer with a blackened bulb to absorb the sun's rays, which should be enclosed in a glass tube from which the air has been expelled, in order that the heat absorbed by the bulb may not be carried off by the currents of air which would otherwise come into contact with it.

By an instrument of this kind I have ascertained, that even in the present wintry month of February the direct solar radiation often raises the temperature to 85°, and even sometimes as high as 105° of Fahrenheit.

A more exact instrument, however, for this purpose is Sir John Herschel's actinometer, in which the amount of solar heat is determined by a coloured liquid inclosed in a wide tube, exposed to the direct rays of the sun for a certain definite space of time, as for one minute in each instance, the dilatation being estimated by the rise of the liquid into a narrow tube connected with the wider one below, when the contents of the latter are expanded by the heat.

Having now considered the most important element of Climate, namely Temperature, in its various relations to the Latitude, Position, and Elevation of the country, I shall next proceed to discuss the other atmospheric influences which affect the condition of plants and animals.

Of these, there can be no doubt, that the one most indispensable to vegetation is the amount of moisture which the air is capable of dispensing to the ground, and that this must be dependent *cæt. par.* upon the quantity present in it, or on its approach to a state of saturation. Now the latter is determined by the instruments called hygrometers, of which there is a great variety, although in constructing them the only two principles which can be regarded as correct, are, either that proposed by Daniell, in which the amount of moisture is estimated by the temperature at which dew begins to be deposited, or that adopted in the apparatus which goes by the name of Mason's Hygrometer, in which the same result is arrived at by observing the difference between two exactly similar thermometers, the one with a dry bulb, the other with one kept constantly moistened, as a means of estimating the rapidity with which water evaporates, which *cæt. par.* will be in proportion to the dryness of the atmosphere at the time being.

In the instrument invented by Daniell we obtain the dew-point directly, by lowering the temperature of the blackened bulb, by means of ether applied to the opposite one, which is covered with muslin; in that of Mason, we are enabled to arrive at the same result indirectly, without the inconvenience of employing ether, by simply comparing the temperature of the dry bulb of the instrument with that of the moistured one beside it.

In either case, we acquire the data for estimating the following particulars:—

1. The elastic force of the aqueous vapour contained in the air.
2. The weight of vapour present in a cubic foot of the air.
3. The weight of vapour required for saturating the same volume.
4. The degree of humidity as compared with complete saturation, which latter is expressed as 1,000.
5. The weight in grains of a cubic foot of air at each reading of the barometer.

All these results are given numerically in Mr. Glaisher's excellent Hygrometrical Tables, published in 1847, and in a more compendious form in Drew's "Practical Meteorology," London, 1855.

Thus let us suppose that we have ascertained by observation, that on a given day, the dry bulb of Mason's hygrometer stood at 64°, and the moistened one at 54°.

Then by reference to Glaisher's Tables we find that the dew-point, if ascertained by Daniell's instrument, would have been 47°, and therefore, the elastic force of vapour 0.337°;—the weight of vapour in a cubic foot of air, 3.75 gr.; the weight of vapour required for saturating this same volume of air, 2.90 gr.; the degree of humidity in the air, as compared to the point of complete saturation, (the latter being expressed by 1,000,) 0.564 gr.; the weight of a cubic foot of air in grains, (barometer being 30°,) 525.700 gr.

Now the humidity of the atmosphere manifests itself to our senses in a palpable manner by the production of three classes of phenomena, namely, 1st, by the generation of fogs and clouds; 2ndly, by the descent of rain and snow; and 3rdly, by the deposition of dew.

A fog is a collection of vapour diffused through the strata of air nearest to the earth, which bears the same relation to the superficial, which a cloud does to the upper regions of the sky.

In either case the phenomenon is occasioned, by the air ceasing to retain the same capacity for moisture which it had before possessed, and by the consequent deposition of the redundant portion in the form of what is called *vesicular vapour*.

The term "vesicular vapour," applied to the diffused moisture which constitutes a cloud or fog, is a theoretical expression originating with the celebrated Saussure, who represented it as composed of an infinite number of little bladders, the walls of which consisted of films of water in the utmost state of tenuity.

But what, it will be asked, is inclosed within these films to prevent their collapsing? If it be aqueous vapour, the specific gravity of each little bladder, with its contents, will be greater than that of the circumambient air, and it would therefore gravitate towards the earth.

It has been suggested, indeed, that as clouds are in a state of electrical tension, the vesicles which constitute them will be kept apart, and prevented from coalescing by their own mutual repulsion.

This, however, although it might retard, would not prevent their gradual tendency downwards, unless the latter were counteracted by some other cause. Fresnel, therefore, supposes the vesicles to be filled with air, which being surrounded by a film of water possessing a certain degree of opacity, and therefore calculated to absorb the solar rays, becomes in consequence hotter and rarer than the circumambient atmosphere; the latter, owing to its transparency, transmitting the sunbeams without becoming heated by them in the same degree. Hence he conceives, that the specific gravity of each little balloon or vesicle may be less than that of the air near the earth's surface, so that it will ascend until it reaches a stratum of the same weight as it is itself.

Whether this ingenious explanation be regarded as satisfactory or not, the term "*vesicular vapour*" may be retained, if we accept, on the authority of M. de Saussure, the fact for which he vouches, that he saw on the slopes of the Alps, on ascending through a region of clouds, a multitude of these vesicles, which he compares to the soap-bubbles blown by children from the mouth of a tobacco-pipe. Considering their extreme tenuity, Gay Lussac conceived, that the constant rise of currents of warm air, which is taking place from the ground, might be sufficient to keep them suspended for a certain time at the elevation at which we find them.

Perhaps, however, a consideration of the phenomena of fog might lead us to frame a more simple explanation with regard to the analogous case of clouds.

The aqueous particles of which the former consist do not appear to be made up of vesicles, but the water is in a state of such minute division, that it overcomes with difficulty the resistance of the air, and therefore gravitates but slowly towards the earth. So long, therefore, as the formation of mist above proceeds with a rapidity equal to that of the deposition of water from below, the fog will continue permanent; and the same thing will happen with a cloud, if we suppose its inferior surface to go on continually wasting, whilst its upper one is in the same ratio increasing.

We need only suppose that the particles of water which form the lower limit of the cloud are taken up by the air at that point, and are thus converted into invisible vapour as fast as they descend to that level.

We must, however, carefully distinguish between the humidity present in the atmosphere in the form of vesicular vapour, and that existing in it in an invisible or aeriform condition.

All liquids have a tendency to pass into vapour, until checked by the pressure of their own *atmosphere;* and this tendency, which goes by the name of the tension of vapour, increases in each case in proportion to the advance of its temperature.

Whenever, therefore, the soil and subsoil are not entirely destitute of humidity, the atmosphere above must contain a certain amount of water, not as vesicular vapour, but in a perfectly aeriform condition.

Now the effects produced upon living beings by the presence of moisture in the atmosphere are entirely different, according as it exists in one or other of these states.

Vesicular vapour, manifesting itself in the form of fog or mist, causes, as every one knows, a sensation of chill, owing to the abstraction of heat from our persons, caused by the moisture which attaches itself to them; and likewise for the same reason interferes with the healthy functions of the skin, and even of the lungs.

But in an aeriform condition the very opposite effect takes place.

Professor Tyndall has lately pointed out, that humid air, or air containing much moisture in a transparent or an aeriform condition, exerts a remarkable influence both in absorbing and in radiating heat. Owing to the former property, aqueous vapour acts as a kind of blanket upon the ground, and contributes in a very striking manner to the retention of its heat.

Hence when the air is perfectly divested of moisture, as in the sandy deserts of Africa, in Siberia, and even in Australia, the cold at night is almost insupportable, owing to the absence of that protection which is afforded by aqueous vapour when present in the atmosphere; whilst during the day the rapid abstraction of moisture from the surface of plants and animals, caused by the dryness, is equally deleterious to both.

And as the radiation of heat from a body is always equal to its power of absorbing it, it follows, that air containing much moisture, will, when it rises into the higher regions, sink rapidly in temperature, in consequence of the heat it sends forth into space; and indeed, according to Tyndall, the amount radiated from air saturated with moisture is 16,000 times as great as that of air perfectly dry.

One cause, therefore, of the profuse rains that occur in the tropics may be the cooling of the heated air, which rises from the earth into the higher regions of the atmosphere, and which, when it arrives there, radiates its heat freely into space, and thus has its capacity for moisture reduced.

Professor Tyndall calculates, that 10 per cent. of the heat radiated from the earth in this country is stopped by 10 ft. of the air which lies nearest to the ground.

It would appear from the recent investigations of M. Duchartre in France, that the refreshing influence of rain and dew upon plants does not arise from the moisture deposited upon their surfaces being absorbed,—for not a particle of water enters into them through the leaves or stem, — but from its supplying the ground with water for the roots to absorb[d]; and likewise, as I should infer, from the check which its presence affords to a too rapid radiation of heat from the parts above ground, by which the prejudicial effects of a too rapid cooling are provided against.

According to the views I have laid before you with respect to the nature of clouds, there will be so much in common between the mode of their formation and that of rain, that it will be better to consider the two together, and thus to enter at once into the inquiry, what are the causes that give rise to a deposition of the aqueous vapour present at all times in the atmosphere, whether this deposition take the form of vesicular vapour or of actual rain-drops.

[d] M. Duchartre found, that the weight of a plant, after being exposed to a night of heavy dew, was not increased, provided only that its roots were so enclosed within a proper case as to prevent moisture from reaching them from without.

He points out three causes for the non-absorption of water by the leaves and stems: 1st, the stratum of air which covers the leaves, and thus intervenes between the plant and the superimposed moisture; 2nd, the waxy covering of the external surface of most parts of a plant; 3rd, the presence of air between the cells of the parenchyma.

But though dew and mist are not absorbed, they contribue to the health of the plant by checking the transpiration of moisture during the night.

The immediate cause in both instances must be a change in the capacity of air for moisture, and this may arise, in the first place, from a direct diminution of its temperature.

Accordingly, the sudden chilling of a body of air may, as I have already pointed out, at any time cause the formation of cloud, or a fall of rain, supposing the atmosphere to be at the time near its point of saturation.

Thus the warm air of the tropics, in passing into a cooler region, would sooner or later reach a point where it could no longer retain the whole of the moisture present in it, and where consequently the latter would be discharged in the form of clouds or rain.

But independently of this, the law regulating the amount of aqueous vapour which can remain in an invisible state suspended in the air, (or in other words, "the tension of aqueous vapour,") involves as a consequence the deposition of water, whenever two currents of air of different temperatures, saturated with humidity, intermix.

A French meteorologist, M. Gasparin, has given a table representing the weight of vapour in grammes present in every cubic metre of air at different temperatures[e], shewing that the difference in the amount of vapour taken up goes on increasing faster than the temperature itself:—

Tension and Weight of Vapour.

Temperature.	Weight of Vapour in the cubic metre of Air.	Difference in each five Degrees of Temperature.
0	5.66	0.00
5	7.77	2.11
10	10.57	2.80
15	14.17	3.60
20	18.77	4.60
25	24.61	5.84
30	31.93	7.32
35	41.13	9.20

Thus let us suppose a cubic metre of air possessing a temperature of 0° to come into contact with an equal quantity possessing a temperature of 30°, and both to be saturated

---

[e] A gramme is about 15½ grains; a metre about 3 ft. 3 in. English.

with aqueous vapour, the two metres, when they mix, will attain the mean temperature of 15°.

Now at this point 2 cubic metres of air take up only $(14.17 \times 2) = 28.34$ grammes of vapour; whereas before they were intermixed,—

The cubic metre at 0° contained . .	5.66
,, ,, at 30° ,, . .	31.93
Together amounting to . .	37.59
So that after deducting . . .	28.34
There will be deposited in clouds or rain	9.25

Now this intermixture of air of different temperatures may be brought about by the interposition of any obstacle to the free transmission of the currents that are constantly circulating around the globe, such, for instance, as a chain of mountains, or even any other more trifling irregularity of surface.

These deflect the stream of warmer air from above, and mix it with the colder below, or *vice versâ*, and in either case, if the strata of air are replete with moisture, the deposition of a portion in the form of rain or cloud will ensue.

Just in the same manner a shoal in the sea causes a change in the direction of the currents of water, bringing up the cold portions to the surface; a fact of which navigators are so well aware, that they are in the habit of foreboding, by a fall in the thermometer, that shallow water is at hand.

Nor is this the only way in which hills contribute to the formation of clouds and rain; for the summits and sides of the former, being cooler than the land below, will condense the moisture of the air brought into contact with it.

Hence in tropical regions, where there are no hills, rain rarely, if ever, falls, because the winds which reach them will generally proceed from a colder quarter, and therefore will contain less moisture than the amount which can be held in a state of vapour, at the higher temperature they will attain on reaching that latitude.

From this cause arises the absence of rain on the great plains of Africa and Arabia, although parallel latitudes in

India and in America, which lie in the neighbourhood of mountainous tracts, receive periodically a much more copious supply of moisture from the heavens, than is the case in the colder parts of the globe.

Indeed, as the capacity of air for moisture increases in proportion to its heat, a greater condensation of it will be occasioned by the intermixture of currents of different temperatures in tropical than in temperate climates, wherever a chain of mountains occurs to intercept and deflect downwards the streams of air from above.

Among the Khasia mountains in Bengal, Dr. Hooker states, that in seven months the rain-fall amounted to 502 inches; and Hartwig sets down that at Guadaloupe at 274.2 French inches; whereas in England, notwithstanding the number of rainy days of which we complain, 45 to 52 inches of rain are regarded as a high average even in the Lake district, the wettest in England. Such appears to be the case at Whitehaven and Cockermouth, places situated near the sea, and therefore affording a fairer representation of the general rainfall in that part of England than the mountains adjoining.

In the latter, indeed, the rain-fall is much greater, though still much inferior to that in the tropical regions alluded to; for at the Head of Borrowdale it is reported as amounting to 151 inches in the year, and at Sprinkling Fell, a mile and a half from Scathwaite, to as much as 211; the exceptional character of which places, however, is shewn, by the enormous difference between these quotations, and those given above with respect to the adjacent towns[f].

Indeed, the average in this island, proverbial as it is for its dampness, is considerably less than in the latter; for at Oxford it amounts to about 26, in London to 25, and in Cambridgeshire to only $22\frac{1}{2}$ inches annually.

The third source from whence the earth derives its humidity is dew, which serves to compensate in some degree for the absence of rain in the hotter regions of the globe.

Although occurring generally in all latitudes, it is most abundant where it is most needed, namely, in the hottest

[f] See Miller's Papers, Ph. Tr. for 1849 and 1850.

and most cloudless regions, owing to a beautiful natural provision, the reason of which cannot be well understood, without entering a little into the consideration of the laws of heat, and the inferences deduced from them by Dr. Wells, in his masterly Essay on the subject of Dew.

It may be shewn, by a very easy and well-known experiment, that all substances are sending forth rays of heat from their surfaces more copiously in proportion to their respective temperatures.

If a heated ball of iron, for instance, be placed in the focus of a parabolic metallic mirror, the heat rays emitted may be rendered palpable, by their effect upon an inflammable body in the focus of a similar reflector placed opposite to it at the distance of many feet.

If we reverse the experiment, and replace the heated iron by a ball of ice, the contrary effect will ensue, for a thermometer situated in the focus of the opposite reflector will then sink perceptibly.

Hence it follows, that an interchange of heat goes on between the two bodies, to the advantage of the one placed in the focus of the second mirror in the first instance, and to its disadvantage in the second.

Applying this experiment to the case of the earth and of the air surrounding it, the former during the day ought to gain by an interchange of calorific rays between itself and the atmosphere, since the latter transmits the beams of the sun, of which it is the first recipient, without absorbing them in their passage. Hence the ground gets hotter in proportion as the day advances.

But during the night the reverse effect takes place, because the earth continues to send forth rays to the air without receiving any adequate compensation. The ground therefore gets gradually cooler from sun-set to sun-rise, so that during this period the difference between its temperature and that of the circumambient atmosphere goes on increasing.

That portion of the air which touches the surface must, however, experience a diminution of temperature by contact with the ground, and will therefore have its capacity for moisture reduced in proportion. Hence will arise a deposition of moisture upon the bodies which lie upon the surface of the soil.

The amount of moisture deposited will, however, vary at different times and seasons, for when the sky is misty, a portion of the heat radiated from the earth will be returned back to it from the opaque body of the clouds, and hence at those times a lesser diminution of the earth's temperature will take place; whereas if it be clear and cloudless, the loss of heat by radiation is not balanced by any counter-radiation from the sky.

Still, however, if a brisk wind should sweep over the ground, the rapid contact of warm air will prevent any great difference between its temperature and that of the atmosphere, so that little or no dew will then form.

But if the air be neither cloudy, nor in rapid motion, a speedy reduction of temperature ought upon the above principles to obtain, and then the deposition of moisture will be proportionately greater.

Dew, however, is not collected indiscriminately upon all bodies,—wool, cotton, silk, and other filamentous substances absorbing it in a great degree; straw, shreds of white paper, grass, and plants of all kinds, still more; whilst glass, chalk, charcoal, sand, and earths in general attract less; and metallic bodies least of all. Now the difference between the temperature of these bodies, and of the air during the time at which dew was formed, appeared to be greatest in those cases where most moisture was deposited; and hence the cause and the effect are seen to correspond. No other reason can be assigned for this difference in temperature, than the greater or less force of radiation belonging to these several bodies, and hence the whole is ultimately reducible to the laws first established by M. Prevost, of Geneva, upon this subject.

It must not, however, be concluded, that a perfectly still atmosphere is the one most favourable to the formation of dew.

On the contrary, a slight movement in the lower strata of the air contributes to its deposition, because by means of it new portions of air are brought in succession into contact with the cooling surfaces of the ground, each of which portions leaves behind it its tribute of moisture.

Thus, then, it appears, that the hottest regions ought to be the ones most favourable to this phenomenon, because the air is there most loaded with moisture, and the ground radiates heat most freely into space, so that the difference between the temperature of the two will be at its maximum in such situations.

Hence it happens, that the dew which falls in the tropics so greatly surpasses in copiousness what we experience in our own climates.

Travellers in tropical countries have indeed reported [g], that at times, when the ground was perfectly dry and the sky clear, the quantity of dew condensed upon the trees adjacent was so great, that it flowed down them like a shower of fine rain, so that it is no wonder, that an exposure to the night air in such regions should be found dangerous to health.

The consideration of the causes which contribute to the humidity of the earth's surface is so intimately connected with that of the winds, that, if for that reason alone, the latter would naturally come under our review in treating of climate.

Thus, for instance, the greater amount of rain which falls in the western quarter of our island, as compared with the eastern, is connected with the greater prevalence of westerly gales, which are more loaded with humidity than any others, and deposit their aqueous contents in the greatest quantity on those portions of land which they first reach.

The cause of this greater prevalence of westerly winds will not be fully understood, until we shall have obtained a clear and complete insight into the laws that regulate those aerial currents, which, independently of local causes, extend at all times over the globe.

These being more regular and uniform than what we experience in temperate regions, present the problem to be solved under its simplest aspect, and hence invite our atten-

---

[g] Boussingault, Econ. Rural., vol. ii. p. 717; Duchartre, Ann. des Sc. Nat., 1861.

tion first, both as holding out to us the fairest chance of being solved, and as promising to afford the readiest clue to the explanation of the remaining phenomena.

Now we can point to a general cause, affecting the entire circumference of the globe, which would bring about an aerial current setting in a certain definite direction on both sides of the equator, as far as from the 25th to the 30th parallel of latitude from either pole.

This cause is the rise in the equatorial regions of heated air, which, as it ascends, will induce a rushing in of the colder and heavier air from higher latitudes to supply its place, thus tending to produce near the surface of the earth a wind from the north above the line, and the same from the south below it, together with a current above flowing in just the contrary direction from the equator towards either pole.

The direction, however, of these aerial currents must be modified by the rotation of the earth upon its axis.

The diurnal movement of the different parts of the globe will be more or less rapid according to their distance from the equator, being null at either pole, but tending from west to east at the rate of 1,000 miles an hour at the line.

The wind, on the contrary, which reaches the latter, will bring with it only the velocity which it had acquired from that part of the earth from which it proceeded, and must therefore, as it travels towards the equator, be moving more slowly in the direction of the east, than the earth is doing in the latter quarter.

Hence its apparent motion will be in the contrary direction, and accordingly the north wind, which is always tending towards the equator, will have a westerly direction impressed upon it by virtue of the movement of the earth upon its axis, or, in other words, be an east wind.

This, perhaps, may be rendered plainer by the following calculation.

In the 30th degree of latitude the earth is said to rotate at the rate of 1,229 feet per second.

This amount of motion is accelerated in the 29th degree by 13 feet, and in the 28th degree by 29 feet per second.

Suppose then a wind to be moving towards the south from 30° to 29°, or from 29° to 28°: if it retain the velocity due to its former position, and has not acquired that of the latter, it will appear to us to be moving towards the west, as its motion east will be 13 feet in the first instance, and 16 in the second, less rapid than that which the earth possesses at this parallel of latitude.

A south-easterly current, therefore, will prevail constantly wherever the colder air from the north is rushing in towards the equator.

It has also been above stated, that whilst the cold air from the Arctic regions is moving towards the south, the warm air of the tropics will have a tendency northwards, forming an aerial current at a higher elevation than the former, which will be moving in a direction exactly opposite to the one before noticed.

This upper current has been perceived, by those who have ascended high mountains, blowing in a direction just the reverse of that taken by the wind in the plain below. Humboldt observed it at the Peak of Teneriffe, and I myself on two occasions on the summit of Mount Etna. It was also proved to exist, from the fact that the ashes ejected by the volcano of St. Vincent in 1812 were wafted to Barbadoes, so as to fall in great quantities on that island.

Now it is well known, that there is a wind blowing at all times about the level of the sea between the two islands, in exactly the opposite direction to the course taken by the ashes.

This south wind becoming colder and heavier as it proceeds, will, by about the time it reaches the 30th parallel of latitude, gravitate to the level of the ocean.

Here, as the velocity it had acquired from the equatorial regions from which it proceeded is greater than that of the region of the globe which it has reached, it will appear to move to the east at a rate proportionate to the difference between its own velocity and that of the earth in this latitude, so as to assume the character of a north-west wind.

Accordingly easterly winds will prevail in the tropics as high as the 28th or 30th parallel, and westerly ones afterwards.

The easterly ones above mentioned are commonly known as the *trade winds*, from the facility they afford to commerce, as the navigator can reckon upon them with so much certainty when he enters the tropics, and can regulate his course by a knowledge of their direction.

They blow as far north as the 30th parallel above the equator, and as far south as the same latitude below it; in the former from south-east to north-west, and in the latter from north-east to south-west; in both cases the primary motion imparted being from either pole, but the tendency towards the west being given to them by their blowing in apparent opposition to the course which the earth is taking in its diurnal rotation.

As, however, the north and south winds will counteract each other's influence near their lines of contact, calms are apt to prevail in the equatorial regions, at least over the sea, where no inequalities of surface exist to disturb the regularity of the aerial currents.

They have been often alluded to by navigators, who dread the long detention they are apt to experience from this cause in latitudes extending from 2° to 12° on either side of the equator, and have furnished a subject for the poet Coleridge in his "Ancient Mariner," where he describes the ill-fated vessel kept stationary in these seas owing to the long-continued calms which there prevail, till thirst and famine had swept away all on board, except the narrator of the catastrophe:—

> "Day after day, day after day
> We stuck, nor breath, nor motion,
> As idle as a painted ship
> Upon a painted ocean."

The south-east and north-west winds will differ materially in their relations to moisture, in consequence of the source from which they arise; as the former, coming from a warmer latitude, will be charged with moisture, which they will deposit in rain as they proceed towards the north; whilst the latter, being derived from a colder region, will become drier in proportion as they advance into a climate warmer than their own, and consequently have their capacity for aqueous vapour increased. The southerly winds also will have a tendency to become more violent as they proceed north, because the diameter of the globe lessens as it recedes from the equator, whilst for the same reason the northerly winds will diminish in force as they advance towards the south.

The trade winds, however, are liable to have their direction changed by local causes.

Thus, as the continent of Mexico is hotter than the sea in the same parallel, a westerly wind will be created, to supply the vacuum caused on the land by the rise of the air over it.

Captain Hall, in Daniell's Essays [h], mentions his having been once thrown out of his reckoning by this circumstance.

In like manner, the monsoons of the Indian Ocean are occasioned by the continent becoming hotter than the sea, and consequently causing the air to rush in from the latter, to supply the place of the more rarified stratum extending over the land. When the sun has its greatest northern declination, the peninsula of Hindostan, the north of India, and China will be the parts most heated, and consequently the monsoon will blow from the south-west. When, on the other hand, the sun goes to the south, the land cools faster than the surrounding seas, and the course of the winds will be to the north-east, producing what is called the northeast monsoon.

Within the tropics the direction of the winds is subject chiefly to the above modifications, which may be accounted for by causes acting periodically, and are therefore capable of being predicted with some degree of certainty.

It is very different with those winds which blow in more temperate regions, in which so many interfering causes operate, that no one can pretend to prognosticate with any confidence the direction which they will assume at any given moment.

Nevertheless above the latitude of 30°, westerly winds, as

[h] Daniell, p. 485.

I have said, predominate [i], being, as Gasparin states, at Paris in the proportion to those from other quarters as follows:—

West	79 } 146
South-west	67
South	63
North	45
North-east	40
North-west	34
East	23 } 46
South-east	23

So that whilst the winds from the west and south-west were in the proportion of 146, those from the east and south-east were only 46, being in the ratio of 76 per cent. of the former to 24 of the latter. Accordingly, whilst the average length of a voyage from Liverpool to New York by a sailing packet is reckoned at forty days, that from New York to Liverpool is only twenty-three. The former is called by sailors going up-hill, and the latter down-hill.

M. Gasparin has given an interesting chart of the average direction of the winds on the continent of Europe.

The general tendency in the northern parts of that tract is, as I have said, towards the west; but this is interfered with towards the south by circumstances connected with the configuration of the land. It will be observed that Europe is crossed by a chain of mountains which, under the name of the Alps and Pyrenees, stretch over it from west to east; whilst to its south extends a long and broad tract of sandy desert, called the Sahara of Africa. The former acts the part of a refrigerator, and gives rise to a cold and heavy stratum of air, ready to descend into the plains below. The latter, by its intense heat, causes an ascending current of air, which flows towards the north, at first at a high elevation, but afterwards descending to the level of the land. Hence northerly winds will predominate in Spain, the south of France, and Italy.

Thus after making due allowance for the irregularities occasioned, by the position of each place, by counter cur-

[i] p. 233.

rents, and by the chilling influence of neighbouring hills, it would appear, says Gasparin, that Europe may be divided into two zones; a northern one, in which south-west winds prevail, and a southern one, in which winds proceeding more or less from the north predominate. The first of these zones is liable to mists and fogs, because it receives the winds from the ocean, but it is also under the influence of the warmth proceeding from the same source, which moderates the rigour of its winters; the second zone, which is brighter, as receiving the dry winds which have swept over the continent, is for this reason cooler than its latitude would denote, and is also subject to greater variations of temperature in accordance with the different seasons of the year.

As the rarefaction of the air will be greatest where its temperature is highest, the winds arising from this cause will be most violent within the tropics, and will take place, whenever the interference of a chain of mountains, or other such cause, produces a sudden change in the temperature of the atmosphere in those regions.

Hence arise the famous typhoons of the China seas, and the hurricanes of the West Indies, the force of which may be estimated by this single fact, that whereas a gentle breeze may be set down as travelling at the rate of about seven miles in the hour, a brisk wind at fourteen, and a gale at forty-one, it has been calculated that the rate of speed in a West Indian hurricane is not less than from 90 to 100 miles an hour [k].

Such a degree of rapidity is sufficient to level trees, unroof houses, and even to prostrate buildings when not very solidly constructed.

In the great hurricane of July 16, 1825, the town of Basseterre, in Guadaloupe, was utterly destroyed. Three

[k] The velocity of these winds may be estimated by the following calculation.

The mean velocity per second of all the winds collectively, taken at Cuxhaven, was estimated at 6.66 metres, the wind which had the highest mean velocity being the north-west, which was as high as 8.70, that which had the least (the east) being 5.58.

This is equal to about 21.85 feet per second, 1,311 per minute, and nearly 15 miles an hour.

twenty-four pounders were blown away, and a piece of deal board thirty-seven inches long, nine inches wide, and seven-eighths of an inch thick, was driven into a palm-tree sixteen inches in diameter.

At St. Thomas's thirty-six vessels were wrecked, and a brig was actually carried up into the air by the whirlwind.

It would appear, however, from Admiral Fitzroy's statements, that even in these latitudes the velocity of the wind has been known to approach that of a hurricane.

Thus on the 25th of October, 1859, the storm in which the "Royal Charter" perished is calculated to have had a velocity of not less than 60, or more than 100 miles, an hour. I should rather accept the former as the nearer approximation to the truth, for the effects of this tempest, terrific as they were, did not approach to those reported in the case of many a West Indian hurricane: but the great storm of 1703, recorded by De Foe, and immortalised by Macaulay's eloquent description[1], certainly seems to have rivalled even the latter in point of intensity.

Now it has been pointed out, that all these great storms possess a revolving as well as a progressive tendency, describing with immense velocity a circle or ellipse round

---

[1] See Macaulay's Sketch of the Life and Writings of Addison. Edinburgh Review. 1843.

"Here it was, that he introduced the famous comparison of Marlborough to an angel guiding the whirlwind. . . . The extraordinary effect which this simile produced when it first appeared, is doubtless to be attributed to a line which most readers now regard as a feeble parenthesis :—

'Such as of late o'er pale Britannia passed.'

Addison spoke not of a storm, but of the storm. The great tempest of November, 1703, the only tempest which in our latitude has equalled the rage of a tropical hurricane, had left a dreadful recollection on the minds of all men. No other tempest was ever in this country the occasion of a Parliamentary address or of a public fast. Whole fleets had been cast away. Large mansions had been blown down. One prelate had been buried beneath the ruins of his own palace. London and Bristol had presented the appearance of cities just sacked. Hundreds of families were still in mourning. The prostrate limbs of large trees, and the ruins of houses, still attested, in all the southern counties, the fury of the blast. The popularity which the simile of the angel enjoyed amongst Addison's contemporaries, has always seemed to us a remarkable instance of the advantage which, in rhetoric and poetry, the particular has over the general."

Fold out

a given area, which, as they proceed onwards, becomes larger and larger, whilst at the same time the rapidity, and consequently the violence, of the movement undergoes a proportional abatement. It has also been observed that the direction is always from right to left above the equator, and from left to right below it.

Having been myself involved in a cyclone of this kind whilst crossing the Atlantic in 1837, as a passenger in the good ship "Mediator," recorded in one of Sir John Reade's charts [m], as one of the vessels which had weathered this tremendous hurricane, and that, I may add, without the loss of a spar, I retain a lively recollection of the fury of the winds, which first assailed us in one quarter, and then, after a temporary lull, re-appeared from exactly the opposite point of the compass, with as much violence as before.

The tempest was ushered in by a singularly fine sunset, a large tract of red light pervading the whole region of the west, whilst the horizon was bordered by a fringe of grass-green, of a tint which I had never before beheld, extending along the margin of the ocean for a considerable distance. This gorgeous spectacle, accompanied as it was by a gentle and favourable breeze, brought all the passengers on deck; but there was something in the appearance of the heavens which, coupled with a fall of the barometer, forewarned our experienced captain of the coming danger, and induced him, to our surprise, to take in sail.

From this wise forethought arose the immunity which we enjoyed, although had our Navigator been aware of the great discovery as to the vorticose movement of these storms which had then been made known, we should have escaped with less peril, as this law would have suggested to him, that in order to get out of the path of the hurricane, he should have steered in just the opposite direction to that which was pursued, so as to avoid encountering the opposite side of the cyclone, after having ridden through the first.

We had thus a striking illustration of the relative advantages of practical skill and of scientific knowledge; as whilst the former enabled our captain to extricate us from the diffi-

[m] See the Chart of the Hurricane of August, 1837.

culties of our position, the latter might have instructed him how to avoid involving us in them at all.

It has also been stated on apparently good authority, that whenever a hurricane rages in a particular direction north of the equator, one of similar intensity occurs at the same time south of the line, blowing in just the opposite one; a fact which would seem to point to some widely pervading cause as influencing both.

I abstain, however, for the sake of my hearers as well as of myself, from attempting the difficult task of explaining these revolving storms, and shall only remark that Mr. Redfield, an American, who has made them his particular study, refers them to the mingling and collision of the upper equatorial current with the lower one proceeding from the poles [n].

If the progressive motion of the hurricane too be, as Mr. Redfield and Sir John Reade contend, a reality, it affords us a means of apprizing distant places, situated in the line which the storm is taking, of its approach, for by telegraphing to this effect from the point at which the storm is first perceived, time may often be given to ships in other ports to prepare for the emergency.

Admiral Fitzroy, however, appears to doubt, whether the storms that occur in these latitudes have this progressive movement combined with the vorticose one attributed to tropical hurricanes, and his method of forecasting the weather, which has on many occasions been of great use in warning seamen of the probability of a storm, is founded upon more

---

[n] Herschel says, (Astronomy, p. 149,) "It seems worth enquiry, whether hurricanes in tropical climates may not arise from portions of the upper currents prematurely diverted downwards before their relative velocity has been sufficiently reduced by friction on, and gradual mixing with, the lower strata; and so dashing upon the earth with that tremendous velocity which gives them their destructive character, and of which hardly any rational account has been given. But it by no means follows that this must always be the case. In general, a rapid transfer, either way, in latitude of any mass of air, which local or temporary causes might carry *above the immediate reach of the earth's surface*, would give a fearful exaggeration to its velocity. Wherever such a mass should strike the earth, a hurricane might arise; and should two such masses encounter in mid-air, a tornado of any degree of intensity on record might easily result from their combination."

general considerations, namely, on the fact of a change in the weather, or on a fall in the barometer, having occurred in some part of the British Islands, as indicating the likelihood of a similar change following elsewhere.

Gentle breezes are useful in vegetation, by keeping plants in slight motion, by strengthening their fibres, and by acting favourably upon the process of their fecundation through the assistance they render in the dispersion of the pollen.

They contribute also to the drying of the surface of the land, and thus in bringing it into a state favourable to cultivation; and they likewise occasion the equalization of the temperature throughout different tracts, by conveying the redundant heat from one quarter to another in which it is deficient; in which way, however, they sometimes injure plants by the sudden abstraction of heat and moisture from them. Dry and hot winds, indeed, are eminently injurious to vegetation, and it is in this way that the Simoom of the desert appears to produce its destructive consequences both upon plants and animals.

It is probable that the poisonous qualities ascribed to this wind are not due to any particular noxious principle, but to its extraordinary dryness, coupled with its high temperature, together with the suffocating effect of the accompanying drifts of sand in which the travellers are enveloped°.

A rapid abstraction of moisture from the surface of the body, as well as from the lungs, may indeed be expected to bring about the most fatal effects; and that this is the main

---

° Columns of sand came moving on.
    Red in the burning ray,
    Like obelisks of fire,
They rushed before the driving wind.
    Vain were all thoughts of flight.
Could they have backed the dromedary then,
    Who in his rapid race
Gives to the tranquil air a drowning force.

    High, high in heaven upcurled,
    The dreadful sand-spouts moved:
Swift as the whirlwind that impelled their way,
    They came towards the travellers.
                    THALABA, bk. iv.

cause, may be inferred from the accounts given of the condition in which the carcases of animals overtaken by the storm are afterwards found, they being described as in a manner mummified, and reduced to that desiccated state to which the corpses of the monks in the Capuchin convent at Palermo are brought by the mere operation of a dry atmosphere, and which indeed may be induced in any animal matter, by placing it under a jar, together with an absorbent substance, such as oil of vitriol or chloride of calcium.

The other circumstances which are to be taken into account in considering the climate of a place or country need not detain us long.

The pressure of the atmosphere varies, as is well known, according to the height of each spot above the level of the sea, and that with so much regularity, that we are able to calculate with the greatest exactness the difference of level between two places by the point at which the barometer stands at each.

The inconvenience of carrying about an instrument so liable to get out of order as the common mercurial barometer, has hitherto very much circumscribed the use of this method of determining heights in the hands of casual observers; but the invention of the *aneroid* has in great measure removed this difficulty, and the reduction of the latter instrument to the size of a watch has really placed it in the power of every person, to note a difference of less than fifty feet in his ascent of a mountain, with the utmost facility.

Unfortunately, these smaller instruments can only be depended upon as far as twenty-five inches, or from 4,000 to 5,000 feet, a range, however, sufficient for any mountain in the British dominions. In the Alps the larger kinds of aneroid must be resorted to by those who wish to avoid the trouble of carrying with them a mercurial barometer, and if, as will often happen, this instrument becomes deranged by the shocks incident upon a long journey, it can be set right by appealing to the smaller instrument, which may be always carried about the person, and therefore is exempt from the same liability to accident.

With regard to the indications of the weather afforded by the rise or fall of the barometer, I shall abstain, for the sake of my hearers as well as of myself, from saying much. In Mr. Leonard Jenyns' "Observations in Meteorology," 1858, may be found (in pp. 130 et seq.) some sensible remarks, on the general principles which determine the relative weight of the atmosphere, and on the cautions that must be exercised in drawing practical deductions from it; whilst in a subsequent chapter (chap. vii.) the general subject of weather-prognostications is fully entered upon.

These, proceeding as they do from one who has for the last twenty years kept an accurate register of the weather, deserve to command confidence; but in my own case, all I should venture to offer on the subject, are two or three general remarks upon the causes which may bring about a fall or a rise in the barometer.

As heat produces rarefaction, a sudden rise of temperature in a distant quarter may affect the weight of the atmosphere over our heads, by producing an aerial current outwards, to supply the place of the lighter air which has moved from its former position; in which case the barometer will fall.

Now such a movement in the atmosphere is likely to bring about an intermixture of currents of air of different temperatures, and from this intermixture, rain, as we have seen, is likely to result.

On the other hand, as cold produces condensation, any sudden fall of temperature causes the column to contract, and sink to a lower level, whilst other air rushes in from above to supply the void; and accordingly the barometer rises.

Should this air, as often happens, proceed from the north, it will contain in general but little moisture; and hence, on reaching a warmer latitude, will take up the vapour of the air, so that dry weather will result.

It is generally observed, that wind causes a fall in the instrument; and, indeed, in those greater movements of the atmosphere which we denominate storms or hurricanes, the depression, as we have seen, is so considerable, as to forewarn the navigator of his impending danger.

It is evident, that a draught of air in any direction must diminish the weight of the column overhead, and consequently cause the barometer to sink.

The connection, therefore, of a sinking barometer with rain is frequently owing to the wind causing an intermixture of the aerial currents, which by their motion diminish the weight of the atmosphere over our heads; whilst a steady rise in the column indicates the absence of any great atmospheric changes in the neighbourhood, and a general exemption from those causes, which, as I have shewn, are apt to bring about a precipitation of aqueous vapour.

As it is possible that the electrical condition of the air may exert some effect upon plants, it may be well to state, what little we know as to the distribution of electricity throughout the globe.

The now familiar fact, that lightning is owing to the discharge of electricity, from a cloud to the earth, or from the earth to a cloud, would prepare us to expect, that the electrical fluid is distributed unequally under different circumstances both in the one and in the other.

It is indeed observed, that during serene weather the earth is generally positive, although its electrical intensity varies every moment, through the influence of passing clouds, puffs of wind, and the like. There is also a variation according to the time of the day, the intensity being greatest about midday, and then decreasing till about two hours before sunset, after which it again increases, and obtains its maximum about two hours after sunset. It then diminishes till sunrise, when it is feeble, but continues to increase till mid-day.

It is also stronger in winter than in summer, and varies in a regular manner during the interval which separates these two seasons.

During violent and transient rains the electricity is enhanced in intensity, but sometimes in the positive, sometimes in the negative direction; and the occurrence of lightning, or the discharge of the electrical fluid from the clouds, which occurs during storms, shews that the electrical equilibrium is then disturbed.

But it would seem doubtful, whether the storm is the cause of the discharge of electricity, or the electricity the cause of the storm; for, on the one hand, we know that the instantaneous precipitation of aqueous vapour disengages a certain amount of electricity; and on the other, that if the particles of aqueous vapour contained in a cloud be in a positive or negative state of tension, they will repel each other, and thereby be prevented from coalescing; whereas when the redundant electricity is discharged, and this impediment to their union is removed, rain is likely to ensue.

After all, our actual knowledge of the electrical conditions of the atmosphere, and of the causes of that condition, may be summed up in a very few words.

Dry atmospheric air, whether it contains much vapour in a gaseous state suspended in it or not, is a non-conductor of electricity; but if it be damp to the feel, or contain much vesicular vapour, it becomes a conductor.

The condensation of moisture, therefore, so as to form a cloud, will create a tendency in the air surrounding to impart its surplus electricity to any contiguous body, whether that body be a cloud or the earth, which happens to be in an opposite electrical condition.

This tendency, however, may for a time be counteracted by an intervening space of cloudless atmosphere, which opposes a non-conducting surface to both the electrically-excited bodies, and during this period the particles of vesicular vapour, being in a similar state of electricity, will be repelled, and thus be prevented from coalescing.

But suppose the intensity of the electricity in either cloud, or in the cloud and the earth, to reach the point at which it can overcome the resistance of the non-conducting medium interposed, and lightning will pass from one to the other, the clouds returning into a state of electrical equilibrium.

Now when in this latter condition, there will be no further impediment to the particles of vesicular vapour coalescing into drops, and rain will consequently ensue.

With the electrical disturbance of the atmosphere the occurrence of hail seems to be in some manner intimately connected, although I am not aware of an explanation of the phenomenon having been offered which is in all respects satisfactory.

M. Peltier imagines, that if two clouds in opposite states of electricity are brought together within striking distance, the particles of vesicular vapour which they respectively contain will be attracted and repelled backwards and forwards for some time before they fall to the ground, and during this interval they will be subject to a rapid evaporation. The latter process, by the cold it occasions, will convert them first into particles of snow; these, however, by being brought into contact with other particles of the same kind, will coalesce, and form masses of ice of greater or less size, according to the time during which they were attracted and repelled, before their ultimate precipitation to the earth in the form of hail[p].

The destructive effect of hail-storms upon vegetation is well known, but it is only in those comparatively rare cases, in which the size of the masses is considerable, that animals are also liable to injury from them. Yet it is stated, that on the 15th of January, 1829, a hailstorm occurred at Cazorla, in Spain, in which the stones weighed two kilogrammes[q] each, and in which they even broke through the roofs of houses.

Although the causes of the electrical disturbances which take place in the atmosphere may as yet not be fully understood, the influence they exert upon vegetation is a matter of general notoriety.

No one can be unaware of the stimulating effect of a thunderstorm in spring and summer upon growing plants, and there can be little doubt that this is connected with the generation of nitric acid in the atmosphere, caused by the direct union of nitrogen and oxygen, which can only by this method be brought about.

This nitric acid easily resolves itself into ammonia, by the

[p] May not Prof. Tyndall's experiments on *regelation*, Sixth Lecture, on "Heat considered as a Mode of Motion," throw some light upon the coalescence of hailstones into large blocks of ice?

[q] A kilogramme is about 2 lbs. 8 oz. English.

substitution of hydrogen for oxygen in the manner which Liebig has explained, and thus affords a supply to plants of that essential ingredient nitrogen, which they appear unable to derive *directly* from the atmosphere, but which they obtain *indirectly* from it through the agency of the electricity thus set in motion.

As these Lectures are professedly limited to the consideration of those meteorological conditions which may be supposed to affect more or less directly the vegetation of a country, I shall pass over various phenomena which come within the range of an ordinary treatise on Climate; but there is one which ought not to be omitted, from its possible connexion with the salubrity of a spot, in relation to animal life, and perhaps in like manner to vegetable. This is the presence in it of that still obscure principle called Ozone.

All that we know for certain of the body which goes by this name may be comprehended in a few sentences.

During the slow combustion of phosphorus, as well as in the passage of a spark from an electrical machine, or during the more gigantic discharges of electricity which take place in nature, a principle appears to be evolved, which physically is distinguished by one only of our organs of sense, namely the smell, from which its name, *ozone*, is derived.

The smell which accompanies the emission of electrical sparks from a machine, and that which manifests itself in cases where a flash of lightning passes within a short distance of the spot where we happen to be, is familiar to most of us, and will be recognised as similar to that taking place during the slow combustion of phosphorus in a jar.

This principle also reveals its existence by certain chemical effects, all of which seem to point in the same direction, namely, as producing in an eminent degree oxygenation.

Thus it changes the sulphuret of lead into sulphate, and the protoxide of manganese into peroxide.

The phenomenon, however, by which its presence can be most readily detected is its power of decomposing the compound which iodine forms with hydrogen, in consequence, as is presumed, of its bringing about the union of the latter with oxygen, and thus displacing the iodine. This is the test originally proposed by Schoenbein, the discoverer of this new body.

Dr. Moffat also has proposed another sort of preparation for the purpose of detecting ozone, but as this is now no longer obtainable, it may be better to confine our consideration to the original papers of Schoenbein, which being all prepared by the same person are more likely to present comparable results.

In these, then, the amount of ozone is estimated, by comparing the colour produced, with a scale of tints marked with numbers from 1 to 10 according to the degree of their intensity.

Accordingly, if ozone be brought into contact with Schoenbein's papers, a blue colour is produced; its intensity being in proportion to the quantity present, just as will happen when a strong acid, like aquafortis, which readily gives out oxygen, is added to the same preparation. It must be evident, however, that this method cannot afford us an accurate measure of the quantity present in the atmosphere, since the amount of this principle brought into contact with the paper will be in direct proportion to the brisker or slower circulation of the air at the time being.

This change to the characteristic colour, arising from the elimination of iodine, and its combination with the starch, takes place, after a few hours' exposure to the air, in certain states of the atmosphere and not in others; and it appears, that even at those times when the prepared papers are deeply affected after exposure to the open air, no change takes place when they are hung up in an inhabited room, so that it seems to be in some measure ascertained, that the purest and most salubrious air is most strongly impregnated with this same oxygenating principle.

A resident practitioner at Weston-super-Mare, Mr. Pooley, in a pamphlet on the causes of the salubrity of that watering-place, assures us, that in the autumn of 1861, when a malignant fever existed in the place and neighbourhood, accompanied with a chill and damp air, no ozone could be detected in the atmosphere; but that when a south wind came on, the disease gradually disappeared, and at the same time a gradual increase took place in the amount of this principle, which on the 21st of November rose to 9, or to nearly the highest point of the scale, at which precise period the type of the disorder changed, and a rapid improvement was noticed.

After these remarks, it may be satisfactory to my hearers to know, not only that ozone seems prevalent at Torquay, but that I have found it generally present in greater abundance here than at Oxford.

With regard to the nature of ozone, it has been conjectured, either that it must contain oxygen, or be itself a modified condition of the same element.

Those who hold the former opinion, regard it either as a deutoxide of hydrogen, or as another combination of hydrogen and oxygen not before isolated.

Those who advocate the second view of its nature, consider it oxygen itself in another form, or in what chemists call *an allotropic* state; and Schoenbein, its discoverer, even conceives, that before oxygen can enter into combination with any body, it first assumes the form of ozone.

It is possible, indeed, that whether arising from the slow combustion of phosphorus, or from the discharge of electricity, its existence may have an electrical origin, and that the first step in the process by which phosphorus becomes oxygenised, consists in the conversion of a portion of the oxygen present into ozone.

But the fact with which we are most concerned, is the presence of this principle at certain times in the atmosphere, in quantities considerable enough to be detected by the test above pointed out. The only connection I have been able to trace between it and the weather is, that at Torquay at least, it seems to abound most during south and south-westerly winds, the average proportion of ozone during their prevalence being 9.5; during east, north-west, west, and south-east winds, about 5.0, or from 5.3 to 4.6; during north winds only 3.1 [r].

---

[r] The proportion may be stated more clearly as follows:—

South and south-west	53.5
North-west, west, south-east, and east	28.2
North	18.3

May we not from this conjecture, that ozone is instrumental in removing by oxygenation offensive and noxious animal products, existing therefore in appreciable quantities in the air, only when no organic matters are present which it can act upon.

It must at the same time be admitted, that we have no direct proof of the effect produced by the air upon the paper in the instances alluded to being necessarily attributable to the presence of ozone, any other oxygenising agent which might be present, and especially nitric acid, being competent to produce the same effects; and as nitric acid in small quantities is generated in the air by electricity, it is conceivable that the effect may be due to this, and not to the hypothetical agent alluded to[s].

Still I cannot agree with Dr. Frankland, who has lately pointed out this possible fallacy, that it is a lost labour to continue observations on the effect produced upon the paper by exposure, for be the principle what it may which produces the coloration, its presence must have a sensible influence upon the purity of the air, by removing from it fœtid and injurious organic effluvia.

It is also quite possible, that this same body may play an important part in regulating the functions of the vegetable kingdom likewise, and although it would be premature at present to speculate upon its specific office, yet for this reason alone it may be well to note the fact of its frequency, in conjunction with the different phases which vegetation assumes, persuaded that no principle can be generally diffused throughout nature, as appears to be the case with this, without having some important and appropriate use assigned for it to fulfil.

[s] This may perhaps explain the remarkable coloration of the papers which often takes place after thunderstorms.

LECTURE III.

CLIMATE shewn to influence vegetation. Monocotyledonous plants are best suited to a hot climate—dicotyledonous best adapted for a temperate one—deep-rooted plants for extremes of heat and cold—plants with shallow roots for equable climates—deciduous trees for climates in which the length of the day is very unequal—evergreens for those in which it is more uniform throughout the year.

Decandolle's five propositions with regard to the adaptation of plants to climate. Influence of the distribution of solar heat upon plants—what kinds are fitted to an excessive—what to an equable climate—what to a bright, what to a cloudy atmosphere—what to a humid, what to a dry climate.

Distinction between wild and cultivated plants with reference to climate.

Farinaceous matter generally distributed throughout the vegetable kingdom.

Plants from which the inhabitants of tropical climates obtain their supply of this material—the Date — Banana—Cassava — Cocoa-nut — Cycas—Sagus—Bread-fruit—Arums—Tacca—Ferns, &c.—Yam—Sweet Potato—Maize—Rice—Millet—Chesnut, &c.— Wheat—Barley—Rye—Oats—Potatoes—Native country of our common cereals—whether derived from other Grasses by natural selection. Comments upon Mr. Darwin's theory on that subject.

HAVING in the two preceding Lectures pointed out the various meteorological influences that affect vegetation, I shall proceed in the present to examine the operation of these combined causes, in bringing about that variety which characterizes the Flora of different parts of the globe.

Flowering plants are divided into those with only one Cotyledon or germinal leaf, and those with two[a]; being accordingly distinguished by the names of Monocotyledonous and Dicotyledonous.

Now these two classes of plants present the most marked differences in their structure, their growth, their mode of flowering, &c.; and from a review of these differences it will be obvious, that whilst dicotyledonous are, as a rule, best adapted for cold climates, monocotyledonous are equally so

[a] Or if with more than two, as in the Coniferæ, having them all in one whorl, or opposite.

for warm ones; so that plants possessing the latter organization may be expected to attain within the limits of the tropics a loftier height and larger dimensions, than they can ever reach within the limits of the temperate zone.

Dicotyledonous plants, such as those which constitute the forests of this and other moderately warm climates, consist of a series of concentric layers of wood and bark, between each of which we may suppose a stratum of confined air to be interposed.

It cannot, therefore, be wondered at that they should be tolerant of cold, both when we consider the slowly conducting power of dry wood of all descriptions, and also that of the air detained within the interstices of the timber itself.

Accordingly, the trees most susceptible of cold are those whose pores are most occupied by sap, the liquid nature of which renders it liable to freeze—an act which, by the expansion it occasions, proves of most fatal consequence at all times to the vegetable organization.

During the severe frost of 1860-61, Professor Balfour states [b], that the bark and wood of several shrubs and trees split open with rents fully half-an-inch wide, and extending eight or ten feet up the stem, and that this rending of the timber was accompanied in one case, at Chatsworth, with a loud noise.

The same thing, indeed, had been remarked by Bobart, at Oxford, during the severe winter of 1683-84.

But monocotyledonous trees, of which Palms afford us the most familiar examples, consist merely of one hard concentric layer of ligneous matter, inclosing a soft pulpy substance, full of juice, and therefore very susceptible of freezing.

Accordingly there are but rare instances of a Palm, or of an arborescent plant of similar structure, existing beyond the limits of the tropics—a few stragglers only, such as the Chamærops humilis, or Dwarf Palm, being found in the south of Europe; the Chamærops excelsa, or Chusan Palm, in China; the Palmetto in the milder parts of the North American States; and the Livistonia inermis, and Seaforthia elegans, indigenous in the south of Australia, and the latter

[b] Ed. Ph. Journ., No. xxvi. p. 259.

introduced as a cultivated plant in as low a latitude as Tasmania.

Nevertheless it must be confessed that the vegetables which afford the largest proportion of farinaceous food to the inhabitants of the cooler portions of the globe, as well as to their cattle, possess the very structure which we have been describing, and indeed in many of the most essential parts of their organization approximate very nearly to the Palms.

Such are the various kinds of Bread-corn, and the Grasses which cover our meadows.

Both these, however, owe their fitness for a colder climate to the circumstance of ripening their seeds in the summer or autumn of the same year in which they have sprouted from the ground, so that, although killed by the severity of the winter, they admit of indefinite propagation from seeds, wherever the summer temperature is such as to allow of the latter coming to maturity.

We may also observe, that of the dicotyledonous trees which belong to temperate regions, those which extend furthest to the north are either protected from cold by numerous layers of bark, as is the case with the Birch, or else are provided with juices not susceptible of freezing, such as the essential oil, which occupies the so-called turpentine-vessels found in the bark and wood of the Coniferæ.

It is also quite natural to expect, that plants with roots that sink deep into the soil should resist the extremes of temperature better, than those which penetrate but a small way beneath the surface. The former derive their nourishment from a depth at which the earth maintains throughout the year an equable temperature; whilst the latter, being affected by all the vicissitudes to which the surface of the ground is exposed, are equally liable to be scorched by the heats of summer, and blighted by the frosts of winter. This may contribute to account for the power which many trees in northern regions possess of resisting cold, inasmuch as they derive their sap from a depth to which changes of atmospheric temperature never extend; and likewise, for the converse reason, it will explain the coolness of the juice which may be extracted from the Palms of tropical regions, where the roots sink deep enough into the ground, to extract their nourishment from parts of the soil which lie beyond the reach of the mid-day heat. Thus there is no cooler or more refreshing beverage in the tropics than the milk of the Cocoa-nut fresh gathered from the tree, the fruit being surrounded by a dense fibrous covering, which prevents the heat of the external air from penetrating into its interior. On the other hand, it is evident that for the same reason, herbaceous plants, whose roots sink very little below the surface, will be ill adapted in general for either extreme of climate, flourishing neither amongst the frosts of the polar regions, nor yet amidst the scorching heats of the tropics.

Being sensible of every change of temperature, such plants can exist only where neither the heat nor the cold is very excessive, and hence are found to abound most within the limits of the temperate zone, diminishing in number and variety both as we advance northward and southward of this medium climate.

Independently, too, of the mere differences in point of heat which distinguish a temperate from a tropical climate, the unequal length of the days in the one, as contrasted with the uniformity in that respect which exists in the other, brings about a corresponding diversity in the character of their respective vegetations.

The short duration of the solar influence during the winter of northern latitudes is well adapted to such plants as shed their leaves in autumn, and which, like certain hybernating animals, fall into a kind of torpor during the colder months; whilst the constancy, both as to the duration and the intensity of the sun's heat, which characterizes the tropics, is propitious to those evergreen trees and shrubs the progress of whose growth is never altogether arrested.

I shall wind up this portion of the subject by stating the few simple propositions, under which the elder Decandolle condensed the whole of what was known in his time, with respect to the conditions, upon which the adaptation of plants to different degrees of temperature is found to depend.

The first of these is, that the power of each entire plant, or part of a plant, to resist extremes of temperature, bears an inverse ratio to the quantity of water it contains.

And hence it follows that the frosts of autumn are less injurious than those of spring, on account of the greater dryness of the wood at the former season.

His second law states, that the power of resisting cold is in a direct ratio to the viscidity of the juices which a plant contains; agreeably to the principle in physics, that water which is thick or muddy freezes less readily than that which is pure or limpid.

This may be one reason why resinous trees, such as some of the Coniferæ, are found to brave so well the cold of the most northern latitudes, and likewise that of the highest mountains of the globe.

A third law is, that the resistance to cold in a plant is in the inverse ratio to the mobility of its juices; just as we find that water may be cooled several degrees below the freezing point without passing into the state of ice, if only kept undisturbed.

A fourth law is, that the larger the diameter of the vessels and cells in a plant may be, the more liable it is to injury from frost; just as we find that water becomes solid much sooner in wide, than it does in capillary tubes.

This is one cause of the tenderness of those plants which consist of an assemblage of cells of considerable diameter, such for instance as the Melon, the Banana, or the Palm tribe.

A fifth law laid down by Decandolle is, that the power of resisting extremes of temperature bears a direct ratio to the quantity of air entangled betwixt the parts of the vegetable tissue.

The down, which covers the exterior of certain organs in many plants, may be intended as a protection against both excessive heat and excessive cold, in consequence of the air contained within its meshes, which serves to prevent the rapid transmission of heat either from without or from within.

This down, accordingly, is found equally amongst the vegetable productions of the tropical and of the arctic regions;

and its presence in the Horse-chesnut may, by protecting the young buds, contribute to the hardiness of that tree, which, although a native of the tropics, flourishes well even in our northern latitudes.

Independently, however, of the amount of solar heat which characterizes particular climates, its distribution over the several seasons exercises a considerable influence upon the growth of plants.

There are certain ones, like the Vine, which require an intense heat in summer to bring their fruit to maturity, but yet are capable of resisting a severe cold during winter; they thrive, for instance, on the borders of the Rhine, or in Switzerland, where the winters are very severe; but scarcely ever ripen their fruit in England, even in Devonshire or Cornwall, where the thermometer rarely falls to the freezing point of water.

Other plants, on the contrary, like the Myrtle, cannot resist cold, but do not demand during any part of the year an exalted temperature.

Thus they luxuriate near the southern coasts of England, but do not shew themselves on the Continent, until we reach a much lower latitude than that of this country.

The former class of plants therefore may be said to be adapted for an excessive climate, and therefore thrive best on continents; the latter for an equable one, and consequently succeed most upon islands.

It may also be remarked, that annuals are adapted for continental climates, as they require heat for the ripening of their seeds, but die away in winter; whilst perennials are better calculated for islands, as it is essential that the winter should not be so rigorous as to destroy them, but not equally necessary that the summers should be always hot enough to ripen their seeds, a failure in this respect for several successive years not entailing the destruction of the species.

It would also seem, that the intensity of solar light exerts an influence upon plants, as it does upon animals, altogether independent of that of heat, many functions of the vegetable economy, such as the so-called respiration, exhalation of moisture, irritability, &c., being attributable to the presence of light, and suspended during the periods of darkness.

Now as different plants require different degrees of stimulus, we may easily understand, why some should thrive best on the tops of mountains, where the light of the sun is vivid, although the temperature may be comparatively low; and others in the bottom of valleys, where the solar influence is more often intercepted by clouds, although the climate itself may be more sultry.

Hence arises the difficulty of cultivating Alpine plants, as for example the Gentians; for although we may be able to command by artificial means almost any required temperature, yet we are unable to select a situation, where the degree of light which such plants demand can be secured to them, unless it be accompanied with an amount of heat which is uncongenial to their constitution.

The different manner in which the stimulus of light affects plants may be illustrated by the various times of the day or night at which they open and shut their flowers, that amount which is conducive to the vigorous discharge of the vegetable functions in one species, being apparently prejudicial to it in another. Thus the Cactus grandiflora, or *night-blowing Cereus*, the most beautiful perhaps, although the most short-lived of flowers, in general begins to open in the evening, but I have once seen it, in cool and dull weather, suspend the expansion of its flower till the morning.

But, independently of the different amount and distribution of the solar influence through the four seasons of the year, other peculiarities of climate may stamp a different character upon the vegetable productions of a country.

Thus two regions, corresponding in all respects in point of temperature, may favour a different class of plants, by reason of the relative degree of dryness which pervades their respective atmospheres.

Ferns, for example, of whatever description they may be, delight in damp, and hence occur most abundantly in insular situations.

Amongst the islands of tropical America, Tree-ferns, together with a numerous assemblage of the herbaceous species, prevail; whilst in the colder climates of European islands those of the latter description are in equal abundance.

Nor are the remote regions of the Antarctic zone, such as the coasts of Australia or New Zealand, less stocked with this class of vegetables.

As the fossil remains of plants found in the Coal formation consist in a far greater degree of Ferns than of any other tribe of plants, we are hence led to conclude, as I have already remarked in a former Lecture, that during the period at which the Coal was in the act of forming, large islands, rather than tracts of continent, prevailed over the surface of the globe, and hence perhaps, as I then explained, may have arisen that genial temperature, and humid atmosphere, which enabled the larger and ligneous species of Ferns to exist in these latitudes, and to be accompanied with Palms, and other tribes, which would seem to indicate a degree of heat not very far removed from tropical.

Cacti, on the contrary, and other succulent plants, are peculiarly suited for arid situations, their spongy texture enabling them to absorb a larger quantity of moisture whenever it is presented to them, and the deficiency of stomata rendering the exhalation less rapid than it is in plants generally, and thereby enabling them to store it up and economise it, during the long intervals of drought, which succeed in those regions to the short and rare visitations of rain and dew.

Hence the thirsty camel finds, within the tough and often thorny skin of the Sempervivums and Mesembryanthemums, a refreshing juice, wherewith to refresh himself in the scorched plains of Arabia.

The difference in the pressure of the atmosphere also affects vegetation in various ways, and especially by influencing the rate of evaporation; and hence it happens, that the plants of mountainous regions, where the barometer is low, differ from those of the places where it stands habitually higher, being in general more aromatic and having a tougher fibre.

Hence the difficulty of cultivating Alpine plants in our gardens, however favourably they may be circumstanced as to exposure to light, humidity, &c.

It may therefore be observed that the conclusions to which we have arrived with respect to the laws of the geographical distribution of plants, are partly physiological, and partly empirical: physiological, when we are able to deduce from the structure and organization of a plant its fitness or unfitness for a particular locality; empirical, when from observing, that plants of a particular structure are ill adapted for supporting certain meteorological conditions,—as for instance excessive heat or cold,—we conclude, that others nearly allied to them are likewise similarly circumstanced.

In a very large proportion of cases, however, both these modes of determining *à priori* the habit of an unknown plant fail us altogether; for it is notorious, that species belonging to the same family, and even to the same genus, are indigenous in parts of the world the most contrasted in point of climate; and that whilst some plants are limited perhaps to one particular locality, others are able to thrive under the most opposite conditions of soil and temperature.

For example, physiology might suggest to us, that arborescent monocotyledonous plants, such as Palms, are best adapted for a tropical climate; and observation might lead us to expect, that if any particular family—such for example as the Melastomaceæ—abound in warm climates, a plant of this or of an allied group would not be likely to grow in such latitudes as our own.

But when we see the Ranunculus tribe equally prevalent in India and in England, we are at a loss to understand in what way climate can have anything to do with its distribution.

On taking a survey of the globe, we find that every extensive region or tract, if isolated either by the sea, by a chain of mountains, or by a sandy desert, from others, possesses a peculiar set of plants, which appear to have spread from one central point round in every direction, until it was stopped by something uncongenial in the climate, or insurmountable in the nature of the obstacles which prevented its further extension.

The reason why these plants, rather than others, have possessed themselves of the regions or tracts in question, lies for the most part beyond our powers even of conjecturing; for although Palms, for example, seem, for the reasons stated, to be best fitted for a tropical climate, yet why one particular Palm should be indigenous in Africa, and another in America, remains altogether a mystery [c].

In like manner the Heaths, with the exception of a very few European species, are crowded together in Southern Africa; whilst in the cognate climate of Australia they are replaced by a nearly allied family, the Epacrises, but true Heaths are unknown.

Thus, too, only one or two Roses are indigenous in America; no Cactuses exist in the Old World, but are represented in kindred climates of Africa by Euphorbias; the Agave, or the so-called American Aloe, is confined to Mexico, but Southern Africa possesses, in the true Aloe, plants of analogous character, though of a distinct family.

Nor can there be any reason assigned from the nature of the climate, why New Holland should abound in trees and shrubs with a sombre light-green foliage, and be so plentifully provided with plants of the Mimosa tribe, which possess no leaves of ordinary construction, but have their petioles so developed and expanded, that they would be taken for ordinary leaves, were it not that their flat surfaces are turned to the right and left, and not horizontally in a manner to receive and intercept the solar light.

These and other peculiarities form the subject of that new branch of botany, which goes by the name of the Geography of Plants—one in itself of great interest, both scientifically and practically.

It ought, however, to be introduced into a course of Lectures, as supplementary to a description of the principal families of plants, since it is only to those who possess some acquaintance with the leading features of the vegetable kingdom, that an account of their distribution over the globe can become intelligible.

[c] In the southern hemisphere Palms with pinnatisect leaves are found, (Areca, Phœnix, &c.); in the northern only those with fan-shaped ones, (Sabal, Chamærops.)

I shall therefore confine myself on this occasion to the consideration of such plants as are cultivated by way of food, either for man or beast, which might naturally be expected to vary in different regions of the globe, in a manner corresponding to the general laws that have been laid down.

It must be observed, however, that an important difference exists betwixt the distribution of wild and of cultivated plants. The former can never establish themselves in a country, where the temperature during any part of the year is too high or too low for the conditions of their existence, and even if either contingency were to take place only once now and then, it would equally have the effect of excluding such productions from its indigenous Flora.

But cultivated plants are quite differently circumstanced in this respect, since if an accident of such a kind were to occur, it is easy to renew the stock by importing fresh seeds or plants from other countries.

Thus, in spite of the warm climate of New Orleans, and of Hyeres in France, it sometimes happens, that a degree of cold sets in, sufficient to destroy the Orange-trees in either locality.

This plant, therefore, can never have been indigenous, though, as many years often elapse without the occurrence of so rigorous a temperature, it is quite within the power of the inhabitants to cultivate it in their gardens, if they find it answer to do so.

At New Orleans, I believe, the proximity to Cuba, where Orange-trees are never liable to that disaster, and afford an abundant produce, renders the people indifferent to the cultivation of this fruit; but at Hyeres, where there is a ready sale for it throughout France, it is well worth while to replace by art the losses, which any such extraordinary cold may have brought about in their Orange plantations.

In short, the number of species which admit of cultivation is greater than that of those which grow wild, inasmuch as the former can be maintained, if during any portion of the year the climate is suitable, the latter only, if during no season it is the reverse.

In this we may trace the operation of a beautiful law of nature, intended to give the widest extension to those kinds of vegetables which happen to possess useful or nutritive qualities.

There is also another provision by means of which the exigencies of the animal kingdom appear to have been consulted, in the laws laid down for the organization of vegetables.

No shoot can be produced from the surface of a tree, no new individual can start into existence from a seed, unless it be furnished with a supply of nourishment in immediate contact with it, and in a condition calculated to enable its delicate organs to take it up.

The substance stored up for this purpose in the seeds, the tubers, the bulbs, and the pith of different plants, much as it may be modified by extraneous matters occasionally super-added, possesses nevertheless in the main a remarkable degree of uniformity, being composed of that material, which, when obtained in a separate form, we call starch, or at least of some body so analogous to it, that chemists would denominate it *isomeric*, but often mixed with a portion of gluten, of sugar, and of some oily matter.

Now it is remarkable that this material, or rather this combination of materials, is exactly the one of all others best adapted for the food of man and many other animals, and hence is emphatically called "*the staff of life.*"

And well does it deserve that name, for whilst it may fairly be doubted, whether any portion of the human race could be brought to subsist for a length of time upon animal food alone, there can be no question, that existence might be prolonged indefinitely by means of that compound of starch, sugar, and gluten, which is present in ordinary flour.

Accordingly, few parts of the world, except those very northern latitudes where vegetation may almost be said to be extinct, are destitute of some indigenous plant yielding an abundant supply of starch; and there are still fewer in which such productions cannot be cultivated by art.

The sources nevertheless from which different nations derive this material are extremely various [d].

In tropical regions, for instance, the plants from whence the inhabitants extract their farinaceous nutriment are more commonly of an arborescent character. Thus in Northern Africa the fruit of the Date Palm supplies food to a large proportion of the human race.

It may appear strange to some of my hearers that in enumerating the plants which supply man with farinaceous matter, I should begin with one which contains in its ripe state no farinaceous matter at all.

But this surprise will not be felt by the chemist, who recollects, that the saccharine matter, which renders the Date so nutritious, is in fact formed by the transmutation of starch into sugar by the processes of the vegetable economy; and that this transmutation, or something very analogous to it, is brought about in the animal, before the starch which we feed upon can be secreted.

In fact, when we subsist upon sugar, nature has already performed for us a part of the task, which the digestive organs have imposed upon them, when our diet is farinaceous; for starch must undergo that amount of chemical change which is necessary to render it soluble, before it can undergo assimilation.

Hence sugar and other analogous, or (as chemists call them) *isomeric* substances, are more easily digested than bodies simply farinaceous, and perhaps for this very reason are better adapted to the more languid temperaments of the natives of tropical regions, as they are to the feeble digestion of infants, who obtain it from their mother's milk. Sugar, as Liebig has shewn, is readily converted into fat in the system of animals, and hence the negroes are said to increase in corpulency, during the period of the year at which the ripe sugar-cane is crushed for the purpose of extracting its juice, and at which the labourers live chiefly upon molasses.

---

[d] See Thouin's *Cours de Culture*, vol. i., for a full list of the different vegetables cultivated for food by man.

It has also been proved* that bees convert into wax, which is a kind of fat, as well as into honey, the sugar which they extract from the nectaries of the flowers they visit.

Hence the saccharine matter of the Date may be pronounced to be highly nutritious, in the sense in which all farinaceous matter is so considered, namely, as furnishing the carbon, by which the heat of the animal body is maintained, and respiration is carried on.

Now this tree demands, according to M. Arago, a mean temperature of 70° Fahr., although from the more exact observations of Rozet, in Algeria, where it flourishes, one only of 62° appears to be sufficient for it.

The latter figure, indeed, appears to represent the mean temperature of Palestine, where, as in ancient days, the Vine and the Date both flourish together; a proof, according to Arago's acute remark, that the climate continues the same as it was at the earliest known historical period, since had it been colder, the Date would not have borne fruit; and if hotter, the Vine would not have succeeded.

There is, I believe, only one place in Europe where the Date ripens into fruit, I mean Elche, in Valentia, for in Portugal and in Sicily, which are somewhat colder than Syria, it seldom or never comes to maturity, and on the coast between Nice and Genoa, which enjoys a mean temperature of 59, the tree maintains itself indeed in a state of vigour, but does not produce Dates. It is, however, cultivated in large quantities in a few favoured spots on this coast, in order to supply Palm leaves for the processions at Rome and other cities in Italy, on the Sunday preceding Easter.

Yet though it abounds throughout that belt of land which stretches along the southern coasts of the Mediterranean, it is not found in truly tropical countries, where the heavy rains which fall during a certain period of the year are as injurious to it, as a low temperature would be. May not this be connected with a peculiarity in this tree, namely, that its male and female blossoms are on different plants? Thus,

---

* See Grundlach's "Natural History of Bees," quoted in Liebig's Treatise on Animal Chemistry, London, 1842.

except in dry states of the atmosphere, the pollen of the male plant will not be wafted to the pistilliferous flower, and hence no fruit will be produced.

This peculiarity of the Date was noticed by the ancients, and is familiar to the natives of Eastern countries, who in their wars on hostile tribes, are in the habit of cutting down the male Date trees, in order to deprive the people whom they invade of their accustomed means of subsistence, although this practice seems prohibited in the Bible[f], and I believe is also so in the Koran.

Accordingly the Date seems to be confined to sub-tropical regions, and its place is taken in the tropical ones of the West Indies, of Africa, and of a large portion of equinoctial America, as well as in the Old World[g], by the Banana; of which there are two cultivated varieties, viz. Musa paradisiaca, commonly called the Plantain, and Musa sapientum, distinguished by its stalk being marked by deep purple streaks and spots.

The succulent stem, rising to the height of fifteen or twenty feet, is formed by the stalks of the leaves which wrap round one another so as to form a kind of sheath, whilst their *laminæ*, or blades, are of an enormous size[h], sometimes as much as ten feet in length and two in breadth, but very fragile, so as to be torn asunder by a gust of wind.

From the centre of the leaves rises a series of purple bracteæ enveloping spikes of flowers, which develop into fruit full of pulpy and saccharine matter, of the size and

---

[f] See Deut. xx. 19, 20. "When thou shalt besiege a city a long time, in making war against it to take it, thou shalt not destroy the trees thereof by forcing an axe against them: for thou mayest eat of them, and thou shalt not cut them down (for the tree of the field is man's life) to employ them in the siege. Only the trees which thou knowest that they be not trees for meat, thou shalt destroy and cut them down; and thou shalt build bulwarks against the city that maketh war with thee, until it be subdued."

[g] M. Decandolle (Geog. Bot., p. 924) inclines to the opinion that it is a native of the Old World, and had been introduced from thence into America.

[h] The specific name of *paradisiaca* was given to the Plantain, from the fancy of the old Botanists, that from the size of its leaves, it was likely to have furnished our first Parents with the material for aprons in Paradise.

shape of cucumbers, and often in their native soil amounting to several hundred.

These fruits afford food to a large portion of the population in the countries in which they are cultivated, being eaten either raw, or cooked in various ways. The fibres of the stem also supply the material for the excellent Manilla hemp. According to Humboldt, a given quantity of ground cultivated with this plant affords forty-four times as much nutritious matter as it would do if laid down with Potatoes, and 133 times as much as with Wheat.

The Banana seems to require a temperature of more than 68° Fahr. in order to ripen well its fruit. It, however, is cultivated with some success in a few of the very warmest localities of Europe, as for instance on the shores of Spain, in Valentia.

In South America an article extensively employed for food is extracted from the roots of the Iatropha Manihot, a plant belonging to the Spurges, or Euphorbiaceæ, which family is characterized by the presence of a milky juice generally more or less acrid and poisonous.

It is remarkable, as containing a perfectly wholesome form of starch, combined with a juice so malignant as to be employed by the Indians for poisoning their arrows. Yet this very juice when expressed may be rendered innocuous by the mere act of boiling, and is then eaten under the name of Cassarine.

After the liquor has been expressed, the starchy matter which remains may be obtained by washing, and after being heated, to destroy any of the poisonous principle that may adhere, forms the nutritious substance called Cassava, which constitutes the food of the inhabitants, and under the name of Tapioca is imported into Europe.

In Jamaica the cultivation of the Cassava is recommended on the score of the readiness with which it grows, the ease with which it is introduced into the ground by cuttings, each of which forms a new plant, the circumstance of its not requiring manure, and the large amount of produce obtained from it.

Another poisonous species of Iatropha, viz. I. curcis, or Physic-nut, having acrid seeds, is nevertheless used in Panama as a culinary vegetable, its leaves after boiling being innocuous.

In the tropical regions of the East Indian Archipelago, the Cocoa-nut is the principal article of daily food, and although probably a native of the Old World, it is very extensively distributed also over the equinoctial portion of the New. A mean temperature of 72°, and an exposure to the breezes and spray of the sea, seem the conditions most favourable to its growth, so that it is not met with at any great distance from the ocean.

Its milk, which consists of oily and saccharine matter, and its kernel, which is a more consolidated form of the same ingredients, with probably a slight intermixture of some nitrogenised element, afford a highly nutritious food to the population, whilst the other parts of the plant, the husky or fibrous covering of the nut, the stem and leaves of the tree, &c., supply materials for almost every other use with which the savage is conversant.

In the islands of the Indian Archipelago, however, another plant, namely a species of Cycas, the circinalis, is extensively cultivated for its farina, on account of the small amount of labour and skill it entails. The stem of the Cycas is swollen into nearly a globular form by the pulpy farinaceous matter which fills its interior. This is scooped out, and constitutes the nutritious material called Sago, so that a whole tree is sacrificed for the sake of the pith which it contained.

The Cycas, however, is not the only material which supplies us with Sago.

Amongst the islands of the Indian Archipelago, and especially near New Guinea, a particular kind of Palm, the Metroxylon, or Sagus Rumphii, which here forms extensive forests, serves the same purpose. In order to procure it, the trunks are cut into logs a few feet in length, their soft interior extracted, pounded, and thrown into water, after which the starch settles at the bottom of the vessel. After being thus separated, it is generally sent to Singapore, where it undergoes a process of refinement.

It is calculated that a tree fifteen years of age will yield from 600 to 800 pounds of this nutritious matter, and that a good-sized tree is sufficient to support a man for a year, whilst the labour of extracting nourishment from it would not occupy more than twenty days, ten days for felling the tree and scooping out the starch, and ten for boiling it, so as to render it fit for food.

In the South Sea Islands the Bread-fruit, Artocarpus incisa, a plant allied to the Fig, takes the place of the other tropical productions above-mentioned, as an article of daily consumption.

It is remarkable, however, that the tree is not indigenous in these islands, being only a variety of a plant called Jack, the Artocarpus integrifolia, which is found native in the Eastern Archipelago.

The fruit may be regarded as a gigantic Mulberry, as in both instances the seeds are inclosed in a pulp, which in the Artocarpus is farinaceous, in the Mulberry succulent. In the Bread-fruit of the Pacific, however, the seeds are developed at the expense of the surrounding cellular substance, and these are the portions eaten.

After roasting them on the fire, they obtain much of the taste of new bread, whereas in the East Indies the whole of the fruit of the other variety, or of the Jack, is eaten, although it constitutes a much less palatable food than the Bread-fruit of the Pacific.

And whilst the fruit of this tree supplies the natives with subsistence, its inner bark furnishes them with clothing, for by a process of soaking, spreading out in layers, which adhere together, and then beating it with a mallet, the bark is converted into a kind of cloth, which when the islands were first visited by Europeans, constituted the principal dress of the natives.

The Bread-fruit is cultivated also in the Brazils, where the mean temperature of the coldest month is 67°, that of the warmest being 81°.

When we recollect that the thickened underground stem (or corm) of our common Arum, or Cuckoo-flower, yields a quantity of starch, which used to be collected in the island of Portland, and sold as Arrow-root, we shall be less surprised at finding that those larger kinds, belonging to the genus Caladium, viz. C. macrorhizum in the Pacific, and the C. esculentum in the West Indies, should supply the inhabitants of those countries with food. The former species constitutes the Eddoes, which the negroes subsist upon; the latter was the principal nutriment of the aborigines of New Zealand before European modes of culture were introduced. It goes by the name of Taro in that island, and, like the tubers of our Potato, has stored up in it for the uses of the vegetable a large amount of farina.

The Arum canariense has lately been imported into Guernsey for the same purpose, and is said to be there perfectly naturalized, requiring only the shelter of a walled garden. If the Report of the Acclimatisation Society can be relied upon, the produce is very remunerative, although I suspect some mistake in the figures given as to the amount said to have been obtained.

In Madagascar, another plant allied in some respects to the Arum family, although differing from the latter very widely in its infloresence, (as it has a perinth containing twelve stamens, instead of a spadix with male and female flowers distinct,) is extensively cultivated for food. It is called the Tacca pinnatifida[i], or Otaheite Arrow-root, being also found in that island, and in other parts of the Pacific. Its tubers yield starch mixed with acrid matter, which is separated from it by washing.

The inhabitants of Australia, and in some cases those of

---

[i] See Herb. Amboinense under the name of T. pinnatifolia; also Hort. Malab., vol. ii. tab. 21. Forster in his Genera Plant. describes it; and Goertner, vol. i. p. 43, exhibits its seed-vessels. The best drawing, however, is in Encycl. Meth., pl. 282.

the Sandwich Islands and of India, are compelled to descend to a still lower tribe of plants for their daily food, namely, to different species of Fern, (such as Diplazium esculentum, Pteris esculenta, Marattia alata, Nephrodium esculentum,) which possess rhizomes sufficiently filled with farina to be taken as food. Notwithstanding their bitter taste, necessity has reconciled the aborigines to their use; just as during the Arctic expeditions the same stern preceptor drove our seamen to assuage the pangs of hunger with the Tripe de Roche, a species of seaweed most repulsive to ordinary palates.

There is also a species of fungus called Melitta australis, employed by the Australians for food, and called by the English Native-bread. It must, however, go through some process of maceration or cooking, before it can be fit for digestion.

But the plant most extensively cultivated in tropical countries on account of its nutritious qualities is the Dioscorea alata, or Yam, the root of which so abounds in fecula, that a single plant will sometimes weigh thirty pounds.

This plant is a climber, belonging to the same family as the common Black Bryony, and nearly allied to the Sarsaparilla tribe. Numerous species of it are found in both hemispheres, but they are confined to the tropics.

The Sweet Potato also, a species of Convolvulus, is much used in these countries. It is a native of India, but admits of cultivation throughout the United States, in Carolina especially, and even extends into the warmer parts of Europe, where it takes the place of the ordinary Potato.

It is a perennial, with trailing stalks putting forth roots at each joint, which swell into large dimensions from the amount of fecula which they contain. It is these roots which afford the supply of nutritious matter for which the Sweet Potato is so much valued.

As we proceed northwards, we find that the plants which supply us with food are chiefly of an herbaceous character, and of annual or biennial growth.

The one, perhaps, which has the widest range of distribution is the Maize, a plant which we owe to the discovery of America; for although it is called Indian Corn, there is every reason to believe it to have been unknown throughout every portion of the ancient world.

Some doubt, I am aware, has been thrown upon this from the report of its seed having been found in connexion with an Egyptian mummy at Thebes; but Decandolle (pp. 946, 947) has exposed the extreme improbability of this being true, as if the plant had been known in Egypt, it would have been depicted on their monuments, as is the case with those kinds of corn which were in actual use amongst them, and would hardly have failed to be extensively cultivated in a country so well adapted for it. Yet although there is every reason to believe that Maize was introduced into Europe from the warmer parts of America, perhaps even from the Brazils, it is cultivated with success in north latitude 50°, near Frankfort-on-the-Maine, and in Gallicia, in latitude 49°. Being an annual, the degree of winter cold is unimportant to it, but a certain intensity of solar heat is requisite for the ripening of its seeds.

Now the mean temperature of Frankfort seems to vary from 65° to 66° of Fahr. This is greater than the mean summer temperature of England, which, except in Hampshire, does not exceed 61° or 62°. We can therefore understand, why Maize is a kind of crop which does not succeed in this country, although it thrives even in Canada, where notwithstanding the rigour of the winter cold, the summers are much hotter than in Great Britain.

But wherever a high summer temperature co-exists with a soil abounding in water, the most profitable kind of culture is that of Rice, owing to the large proportion of farina which it contains.

A mean summer heat of about 73° seems to be that which suits it best, but it is cultivated in Lombardy, in the south of France, and in Hungary, where the heat is several degrees inferior.

Notwithstanding that rice cultivation seems to be circumscribed within narrow limits by the joint conditions of high temperature and great humidity, it would appear that a larger proportion of the human race subsists upon this than upon any other description of food, although its dependence upon humidity renders it a precarious crop, and exposes the people who trust to it for subsistence to frequent scarcities and famines.

In drier situations, indeed, where the climate is warmer than our own, some of the numerous species of Millet, viz. Panicum miliaceum, P. italicum, &c., and a species of a grass called Holcus, which goes by the name of Negro Corn in Africa, and by that of Sorghum in Italy, is cultivated in many parts of Asia, of Africa, and of Italy. The latter is not only used as food, but an intoxicating liquor is prepared from it by fermentation, which goes in Nubia by the name of Dhourra.

In Tuscany, where the soil and temperature are eminently favourable to the growth of the Chesnut, the fruit of that tree, instead of being merely regarded as an agreeable addition to the dessert, is used by the peasants as a substantial, and indeed a principal, article of daily consumption. The Chesnuts are not merely roasted as with us, but also ground into a fine meal, and in that condition are cooked in a variety of ways, especially forming a kind of pudding or Polenta, when mixed up with a quantity of Olive-oil sufficient to give it a proper consistency.

I will next proceed to enumerate the plants which furnish nourishment to the inhabitants of this and other similarly circumstanced countries.

Omitting the Quinoa, (Chenopodium Quinoa,) a species of goosefoot confined to the high plateaus of Southern Peru, but probably capable of introduction into Europe, and the Amaranthus fariniferus, cultivated in the same manner on the table-lands of the Himalayas, together with certain tuberous roots used for food in parts of the globe rather warmer than our own, such as those of the Oxalis tuberosa on the Cordilleras of the New World, and of the Sagittaria sagittata in China, I shall notice first Wheat, as being the cereal which extends over the widest range of latitude, although its produce varies greatly according to latitude, in cold countries not exceeding 6 or 8 fold, in the north of Mexico 17, and in the southern part of that country 24 and even 35 fold that of the grain sown.

The power which this plant possesses of resisting cold in northern regions, and of enduring a high temperature in tropical ones, may perhaps be connected with the great depth to which its roots, comparatively to the size of the plant, are able to penetrate into the ground, sometimes, it is said, in light and yielding soils extending as far as six feet from the surface.

This circumstance enables it to draw its sap from a depth at which the earth is, comparatively speaking, little under the influence of the vicissitudes of diurnal temperature.

It is cultivated indeed, according to Humboldt, near Caraccas, at a height of 1,600 feet above the sea; in Cuba at even a lower level than this; and in the Mauritius almost to the water's edge.

It is also grown in the Island of Luçon, and in many parts of India, as well as near Canton, by sowing it at that period of the year when the temperature is lowest. Thus in Bengal, wheat, barley, oats, beans, &c., are sown in October, and reaped in March or April.

It would seem that the extreme point which admits of the cultivation of Wheat is about 71° Fahr.: now the winter climate of Havanna does not exceed that point, and that of Canton and Mexico rises only to 65°.

In Egypt, where Wheat is grown abundantly, the mean temperature of the three coldest months is as low as 58°, so that the seed is sown in December and the crop is reaped in February.

Such are the extreme boundaries of the cultivation of Wheat in a southern direction.

On the other hand, a summer temperature of 58° is generally set down as requisite for the ripening of this vegetable, so that it is limited in its extension northwards by this circumstance.

Accordingly in this country, its cultivation does not succeed at a height of more than 600 feet above the level of the sea, and is limited in point of latitude to the neighbourhood of Aberdeen, situated in the 57th parallel.

It is not long ago, indeed, namely in 1727, that a crop of Wheat in Edinburgh was noted as a curiosity; nor was its cultivation at all extensive throughout Scotland even so late as 1770.

At present, however, abundant crops are to be seen in the Lowlands, and its culture is pushed as far as the Moray Firth. Now as the mean temperature in summer of those parts of Scotland does not exceed 57°, or even 56°, we must account for the growth of Wheat by the length of the day, which seems in some measure to compensate for the defect of solar radiation—agreeably to the statement of Boussingault, that Wheat requires 8,248° of Fahr. to bring its grain to maturity, so that a longer duration of solar heat will produce the same effect as a shorter period of greater intensity.

It appears from an interesting report on the climate of Scotland, published in 1862 by the Meteorological Society of that country, that the mean temperature of the whole island varies very little, being 47° 1' for the east coast, 47° 2' for the inland portions, and 47° 8' for the west; the only places which fall below this standard being the Orkney and Shetland Islands, and the remoter Hebrides, where the mean temperature is only 45° 8'.

But though the temperature of the year may be the same throughout Scotland, its distribution throughout the different seasons differs; and hence it happens, that owing to their lower summer heat the western parts of the island scarcely admit of the successful cultivation of Wheat, and that elsewhere it is limited to a height of 500 feet above the sea in the south, and 100 in the extreme north.

It is curious, as corroborating the views of M. Boussingault, that the ripening of Wheat in the northern part of Scotland took place at a somewhat lower temperature than in the south of that island, owing to the greater length of the days in the former.

The following table represents the temperature required to mature crops of Wheat and Barley in Scotland:—

		DEG.	DAYS.
Colloden, N.	Wheat	8188	156
	Barley	6560	119
	Oats	6767	123
East Linton, S.	Wheat	8362	159
	Barley	6900	129
	Oats	7125	133

In Norway Wheat is cultivated as high as Drontheim, in lat. 59°; in Sweden up to the 63rd parallel; and in Russia it is met with extending to St. Petersburgh, in lat. 59° 5′, where the summer heat indeed is said to average 60°.

Now this represents very nearly the mean temperature of that season in the midland counties of England, nor does Penzance even, mild as it is in winter, exceed that point in summer. On the Hampshire coast alone the thermometer is quoted nearly as high as 63°. As the summers of this country in general so little exceed the point necessary for the successful cultivation of Wheat, we can understand why the western side of the island, which is cooler in summer than the eastern, should be better adapted for the growth of Grass than of Corn. And hence we perceive why Wheat is carried from the eastern to the western counties, whilst cattle are driven from the eastern to the western.

On turning to a map of England it will be found, that all our principal corn districts are situated on the eastern side of the island, from the Lothians to Kent.

In the western part of Scotland, indeed, the summer sun is so insufficient for the ripening of Wheat, that other kinds of produce take its place altogether.

The other species of cereal Grasses cultivated in this country are indeed exempted from the influence of the winter's cold, by being sown after the rigour of the season has passed away. Hence it is not surprising, that Barley should extend further to the north than Wheat, as being sown in March, and accordingly we find the hardier kind, called Bere, at the extreme limits of Scotland, as likewise in the Orkney and Faroe Islands, in Lapland, at the North Cape in latitude 70°, in Russia as far north as Archangel, and even in Siberia in lat. 58° and 59°.

Hence it seems to admit of being cultivated, wherever the summer temperature does not fall short of 46° or 47°, although this has been of late disputed by an author[k], who contends, that at its northern limit in Norway the mean temperature in summer is 53° 4′.

He alleges, moreover, that in Iceland, where the mean temperature is 49° 5′, Barley will not ripen; but this I imagine to arise, less from the low mean of the thermometer, than from the unseasonable rains that ravage that desolate island during the period of its ripening.

In a southern direction the limit of Barley is considerably less extended than wheat, as it is unable to endure an equal intensity of solar heat.

Rye would seem to be rather less hardy than Barley, for it ceases to grow in Sweden about the 66th parallel of latitude, and in Norway at about the 67th.

The northern boundary of the cultivation of Oats is not yet ascertained with accuracy, but this cereal extends at least as far as the most northern point of Scotland, and is therefore but little, if at all, less hardy than rye or barley.

The relation as to temperature subsisting between these four kinds of cereals is also clearly discerned, by comparing, one with the other, the point above the level of the sea, at which their respective cultivation has been ascertained to cease.

In Switzerland,

Wheat may be cultivated as high as 3,400 feet.
Oats . . . . . 3,500 „
Rye . . . . . 4,600 „
Barley . . . . . 4,800 „

Where the only apparently anomalous fact is, that Oats should not admit of being grown at about the same elevation as Rye and Barley.

Of all known vegetables, however, the Potato is the one which has the widest range of distribution.

[k] Whitby, in Roy. Agr. Journ., vol. ii. p. 38.

It is supposed to be derived from the neighbourhood of Lima, or from Chili, and yet it has been diffused throughout the whole of Europe, and even attains a higher latitude than barley itself.

It succeeds in Iceland, where, as above stated, no kind of cereal can be made to grow; and it is remarkable, that whilst it yields an abundance of wholesome nourishment in a cold climate, it degenerates in countries whose mean temperature approaches to that of the regions of which it appears to be a native. In the latter, indeed, it is a poor stunted production, with tubers so small as to yield but little nourishment. Art has brought about that development in them which renders the plant serviceable to our uses; but in so doing, it may be presumed to have induced in it an unnatural condition, and thereby rendered it liable to those diseases which have of late so much detracted from its utility.

It is, however, by no means a peculiarity of the Potato, to be derived from a country considerably warmer than those in which it is principally cultivated.

All our staple productions may be reasonably suspected of having had a similar origin.

It is true, that the native country of the cereal Grasses is still open to discussion. Wheat and Barley have, it is said, been found growing wild, in Persia, Mesopotamia, and on the banks of the Euphrates; and a writer in the "Edinburgh Philosophical Journal" in 1827, comes to the conclusion, that the valley of the Jordan may be the native country of all the cereals.

But who does not perceive, that the existence of wild Corn in a country formerly inhabited by an agricultural population, only implies that the climate is warm enough to allow of its maintaining itself when once introduced, not that it was originally itself a native of the soil.

We know, in fact, nothing as to the origin of Oats or of Rye, and with respect to Barley, we can scarcely infer its native country, from being told that it grows wild in countries so remote from each other as Tartary and Sicily.

Wheat is said to be a native of Sicily, which was the fabled birthplace of Ceres; but then we are told, on the other hand, by Strabo, that it came from the borders of the Indus.

It is, however, certain, that in our northern latitudes none of the above cereal Grasses will diffuse themselves spontaneously, for although a few scattered seeds may come up of themselves from a field which has been the year before in a state of cultivation, yet it seems generally admitted, that there is no instance of these crops maintaining themselves for any length of time in ground abandoned to itself.

This therefore would seem at least to indicate, that our cereal Grasses have been derived from a warmer climate, unless, indeed, we feel disposed to adopt the views of Darwin, and imagine them to be produced by a long process of natural selection from other kinds of Grasses.

The experiments of M. Favre, who professes to have converted the grass called Ægilops ovata gradually into Wheat, by cultivating the former for a number of successive years in a rich soil, have excited much attention, and if their accuracy were admitted, the conclusion we should arrive at would be favourable to the hypothesis of Darwin; but another Frenchman, M. Goudin, maintains, that the changes which Favre had remarked in the Ægilops arose from a crossing that occurred between this grass and the contiguous wheat. Hence would occur hybrids, which partook of the characters of both the parents, and which are not permanent, as, according to M. Favre's view, they ought to be, but revert gradually to the original stock.

I do not, therefore, venture to bring forward the case of the Ægilops as affording any independent support to the doctrine of Darwin regarding the gradual transmutation of species, although those who are already persuaded of the truth of that hypothesis, may feel themselves justified in interpreting the facts observed by M. Favre in accordance with it.

We do not, however, find in other cases any such tendency to change, even after a period of immense duration.

The Spruce Fir, which Dr. Heer describes in the upper pleiocene beds, near Happisburgh, in Norfolk, making part of a submarine forest, has all the characters of the Spruce Fir of the present day; and the most extreme examples of divergence from the normal condition of a species which can be brought about, are all, so far as we know, capable of breeding together, and of producing a fertile offspring.

I am of opinion, indeed, that divines have presumed too far upon our imperfect acquaintance with the laws of creation, and of the limits set to their operation, when they pronounce dogmatically, that the introduction of a new species of animal or plant necessarily requires the immediate interposition of the Divine energy; considering that other phenomena equally mysterious and unaccountable are attributed without scruple, or offence taken, to the agency of secondary causes.

But on the other hand, whilst recognising the existence of provisions, which seem intended for the adaptation of each species to the new circumstances under which it might be placed, and for the production of varieties suited for these altered conditions, we must not ignore the operation of another law, which those who at the earliest times meditated upon physical phenomena pointed out, that, namely, apparently framed for the maintenance of the order of creation instituted upon our globe, which, like that of the planetary systems in general, seems to be secured by means of *constant change*, oscillating within certain definite limits.

Amongst organic bodies, as I have observed on another occasion [1], this stability appears to have been provided for by means of which we have some comprehension, namely, by the limit which nature has imposed upon the divisibility of matter; by which simple law, as an ancient poet long ago observed, the immutability of the Universe is effectually secured:—

"For if o'ercome
By aught of foreign force, those seeds could change,
All would be doubtful; nor the mind conceive
What might exist, or what might never live [m]."
                    GOOD'S LUCRETIUS, bk. i. 639.

[1] See Introduction to Popular Geography of Plants. (Lovell Reeve, 1855.)

[m] . . . . . Nam, si primordia rerum
Commutari aliquâ possent ratione revicta,
Incertum quoque jam constet, quid possit oriri,
Quid nequeat; finita potestas denique quoique
Quâ nam sit ratione, atque alte terminus hœrens;

Yet the elements themselves, no less than their combinations, seem susceptible of certain modifications in their properties, arising perhaps out of a new arrangement of their constituent particles, as the phenomena of *allotropism*, lately made known to us by chemists, seem to demonstrate.

Amongst organic bodies, indeed, the *machinery* is veiled in greater obscurity, but the *end* is still the same; for here also Nature appears to have attached equal importance to the preservation, unchanged throughout all time, of that order and arrangement which she has herself instituted.

Hence, apparently, the impediments she has set up to the production of varieties by hybridization, impediments, the existence of which we must all acknowledge, however much we may be in the dark as to the causes which produce them.

These remarks of mine were introduced into a popular work before the views of Mr. Darwin had been promulgated; and I am still inclined to think, that there is no real inconsistency between the law therein enunciated, and all that observation has as yet substantiated, with respect to the effects of natural selection in inducing new and widely divergent varieties. For whilst the great array of facts so ingeniously brought to bear upon his theory by Mr. Darwin, compels us to grant, that a much wider range of variation must be allowed to species than had been hitherto contemplated, it is still open to us to pause, before we go to the length of maintaining, that this power of variation has no limits prescribed to it—at least until some undisputed instance shall have been adduced of varieties gradually merging into species[n]; or, in other words, so far divergent in fun-

Nec totiens possent generatim sæcla referre
Naturam, motus, victum, moresque, parentum.—Lib. i. 585.

[n] The following instances, which Mr. Darwin has adduced in support of this position, do not appear to me conclusive.

Two varieties of Maize, a dwarf and a tall kind, did not cross.

Three varieties of Gourd became less fertile in proportion to the differences between them.

Yellow and white Verbascum, when intercrossed, produced less seed than each did separately.

Tobacco plant is less fertile with some varieties than with others.

damental points from the normal type, as to be incapable of breeding with other members of the species from which they originated, and likewise exhibiting no tendency to revert to the characters of the primary stock.

I should moreover be inclined to withhold my assent to these views, until some of those vast gaps have been bridged over, which at present interrupt the chain of connection between one part of the system of organic nature and that to which it most nearly approximates, and which, so long as they exist, present in the minds of many a formidable obstacle to the entire reception of the ingenious and fascinating theory of Natural Selection.

LECTURE IV.

Power of man to modify climate—by cutting down timber. Instances shewing the effect of this—in America—in other countries. Causes of the change of climate producible by man's agency—whether Europe is warmer now than formerly—Iceland and Greenland seem to indicate the contrary. Limited power of man to modify climate as compared with natural agencies—the latter, however, have never, within the limits of our experience, operated so banefully upon our well-being, as our own bad passions have done. Power of man to acclimatize plants considered—analogous case in the animal kingdom. Modes in which varieties may arise—Archbishop Whately's scheme for introducing hardier varieties of the plants of warmer countries. Other modes of adapting plants to climate—Street's directions—Root pruning. Description of plants best fitted for introduction into a colder climate. Adanson's principle as to the sum of heat necessary for the flowering of each plant. Exceptional cases of climate as affecting plants considered—Cornwall—Tierra del Fuego. Practical bearing of these principles upon farming. Lawes on the influence of climate upon the crops. Cultivation in England—of Maize—of Hops—of the Vine. Daniell's suggestions. Selection of soil and situation for plants. Orchard houses. Gardens under glass. Importance of meteorological records for determining the exact character of the climate in each locality. Combination of circumstances rendering Torquay suitable as a winter residence for invalids. Conclusion.

HAVING in my preceding Lecture pointed out, in what manner, and to what extent, the distribution of plants over the globe is affected by climatic influences, and also the kinds of vegetables which are found to be most suitable for cultivation in each region, I shall proceed in the next place to consider, how far human agency is capable of modifying these natural arrangements, either by the change it may induce in the character of the climate itself, or in that of the plants subjected to its influence.

The former of these enquiries belongs more properly to the subject of Husbandry, the latter to that of Horticulture, although it must be admitted, that both these departments of Rural Economy are to a certain extent involved in either branch of the enquiry.

In considering the power of man to modify the climate of

the country in which he resides, it may be observed, that the quantity of rain which falls in a particular district seems to be often affected by the cutting down of its forests, and likewise that the water which the soil receives from the heavens will be got rid of more rapidly after a clearage of its timber.

Trees will naturally arrest in their progress the clouds and mists which sweep over the country, and cause them to deposit their moisture upon them, whilst at the same time evaporation from the surface will go on more rapidly, when the wind has a free circulation, than when the ground is protected from it by wood.

Of the extent to which these causes operate, Boussingault has given a striking example, in his account of the change produced in the climate of Venezuela by the cutting down of the forests, in regard to the fall of rain, and the humidity of the district. The Lake of Valentia in that province, being destitute itself of an outlet, is calculated to gauge with the greatest nicety, by the rise and fall of its waters, the increase or diminution in the rivers that discharge themselves into it.

During the early part of this century, when Humboldt visited the spot, the lake was reported to him to be constantly lessening. By the recession of its waters, low islands formerly standing just above the water's level became converted into hillocks, land once covered by water had been transformed into beautiful plantations of Bananas and Sugar-canes, and a bed of fine sand intermixed with fresh-water shells was detected several yards above the level of the lake.

This diminution in the rivers of the district was traced to the felling of timber on the contiguous hills, followed by the falling off of rain; and a confirmation of this theory was afforded by the subsequent increase in the dimensions of the lake, which was observed by M. Boussingault twenty-five years afterwards, owing to the partial return to a state of nature, which followed upon the desolation caused by the civil wars during the struggle for independence. Hence as timber was no longer felled to the same extent, rain fell in greater abundance than before.

Another lake without an outlet, situated in New Granada,

supplied Boussingault with a second and a similar instance of the connexion between the quantity of timber and the amount of rain. Here the recession of the waters was a matter of general notoriety, and coincident with this diminution had been the clearing of the surrounding forests, to afford fuel for the salt works that exist in the neighbourhood.

Nor can this have arisen from any change of climate, for in other places in the same neighbourhood, where no clearings have taken place, and the country continued to be left to nature, the level of the lakes had undergone no change since the memory of man.

Humboldt in his travels through Siberia was assured, that connected with an increased cultivation of the soil in those regions, the same diminution had occurred in the depth of their lakes; and those of Neuchatel, Bienne, and Morat, must have sunk since Switzerland began to be peopled, if it be true that they formerly constituted one continuous sheet of water.

Saussure also asserts that the same diminution has taken place in the Lake of Geneva.

Berghaus has pointed out, that the Oder and the Elbe are both inferred to be undergoing diminution, from observations made from 1778 to 1838 with respect to the former, and from 1728 to 1836 on the latter river.

Whether, however, this has arisen from a decrease in the amount of rain, or from an increase in the rate of evaporation caused by the clearance of the country, may admit of doubt.

Gasparin shews, that during the last century the quantity of rain that falls annually has remained stationary, at Paris, at Milan, and in other places; whilst Boussingault states, that in the province of Popayan, where the supply of water had fallen off so considerably that the mines of Marmato suffered, the rain-fall nevertheless had not diminished.

An interesting confirmation of this truth is afforded by the Island of Ascension.

In this case it appears, not only that the removal of the trees was followed by the drying up of the only spring which the Island possessed, but also that the restoration of the timber brought with it the recovery of the lost water.

Boussingault, however, contends, that although this may be conceived to have happened from the timber having impeded evaporation, and thereby preserved the water that fell from the heavens, it could not have arisen from any decrease in the actual rain-fall effected over so small an area as that which the Island of Ascension occupies.

It is, however, enough for our purpose if we are able to substantiate the fact, that by the removal of the forests we have it in our power to modify the character of the country with respect to humidity, whether this be brought about in one way or in the other; of which fact I apprehend there is abundance of proof.

It is not many months ago, that a correspondent in the "Times" newspaper predicted the gradual decay of Great Britain from the exhaustion of the vegetable mould which imparts fertility to our fields, and his remarks were considered as of so much weight, that the Editor thought it necessary in the subsequent number to devote a leading article to the subject.

That vegetable mould supplies plants directly with nutriment, or that it can ever be exhausted, so long as the crops are maintained in a state of luxuriance by tillage and manure, is a notion which I had believed to have been altogether exploded by Liebig; but that many of those countries, which in ancient times were fertile enough to maintain a large population, are barren and desolate at present, is nevertheless a fact which does not admit of dispute.

To me the cause of this deterioration appears obvious, as arising from the denuded state of these countries as regards timber, for which we need not go further than to many of the islands of the Archipelago, to parts of Greece, and even of Italy.

But that the land can be maintained in a state of fertility for an indefinite period, where the country enjoys a sufficient degree of humidity, and where the natural richness of soil supplies the mineral ingredients which the crops require in sufficient abundance, in those cases in which they are not furnished from extraneous sources, is evident, I think, not merely from the case of Egypt, where the waters of the Nile convey an unusual supply of nutritious matter from the high lands of Ethiopia, but also from that of Naples, Tuscany, and even of parts of Sicily, and indeed wherever the ground obtains fair play from the industry of man.

Where the reverse is the case, I should attribute it rather to the aridity caused by the destruction of the forests, than to the exhaustion of the vegetable matter itself.

In addition to the facts already adduced in proof of the influence of trees upon the fall of rain, it may be mentioned, that Lower Egypt, which is usually cited as a country where rain never falls, has lost this character, having, as it is said, experienced of late showers occasionally of rather a heavy description, at least in the neighbourhood of Cairo and Alexandria; this remarkable change being coincident with the planting of trees, which the late Pasha has brought about to a considerable extent in the neighbourhood of his capital, and in other parts of Lower Egypt.

It would appear, then, that man is really capable of exercising a certain control over the humidity of the climate, by thinning the forests, or by renewing them in the manner represented; nor can it be doubted, that the same effect will be brought about by drainage, which carries off the redundant waters into their appropriate channels, instead of allowing them to stagnate upon the surface.

And in thus altering the character of a country with respect to its humidity, he may hope to bring about a corresponding change also in its temperature, for the tendency of swamps and stagnant waters is to cool down by their evaporation the surface of the earth, as well as to intercept the rays of the sun by the mists and fogs they engender.

Nor must we forget the genial influence, upon a soil properly drained, of summer showers, possessing as they often do a temperature from 10° to 20° higher than the ground. This is lost in an undrained soil, because the water which descends from the heavens is mixed with the moisture of the ground, and thus adds very little to its warmth.

It has been calculated by Mr. Raikes, from experiments made at Chat-Moss, that the temperature of the soil when drained averages 10° more than it does when undrained; and this is not surprising, when we find that 1 lb. of water evaporated from 1,000 lb. of soil will depress the whole by 10°, owing to the latent heat which it absorbs in its conversion into steam.

A similar influence in depressing temperature will be exerted upon a newly-peopled country by the forests which extend so generally over its surface, at least by intercepting the sun's rays, if not, as we have endeavoured to shew, by increasing the rain-fall. Would not this circumstance lend probability to the idea, that countries like our own were formerly colder than they are at present; not that any secular variation of climate can be supposed to have taken place within the limits of human experience, but that the general character of the country, whilst yet unreclaimed and overspread with forests, would be naturally more humid, and consequently colder.

It has been already remarked that Wheat was but little cultivated in Great Britain some centuries back; might not this have been owing to the climate being formerly of too low temperature to admit of its general introduction?

But this opens a large question, namely, what was the temperature of the ancient world as compared to that of the modern, and of England in olden times in relation to what it is at present?

In entering upon this enquiry I have no intention to go back so far as the Glacier period of Agassiz, or even the age of Flint Instruments, of which we have lately heard so much.

Suffice it to say, that even the latter epoch, which is considered to have been vastly more modern than the glacial period, which overspread Europe with those boulders, that have been carried to such a distance from the rock from which they originated, is supposed by Sir Charles Lyell to have possessed a climate much colder than the present, so that at the time when the inhabitants of what is now the valley of the Somme, in Normandy, chipped into the shape of arrowheads the flints of the district, the river was frozen over every year for several months, and floating masses of ice deposited in the bed of the river, as they melted, blocks of stone and strata of mud and sand. Hence those fragments of hard and angular sandstone which occur both in the lower and higher gravels near Amiens, and hence those irregularities in the strata which have been remarked in the same formations.

Whether these indications of a continuance of the cold of the glacial epoch to these comparatively later times lend any probability to the notion, that during any portion of the historical period the climate of Europe was more rigorous than it is at present, I shall leave for others to determine—strictly confining myself to the enquiry, whether there is any evidence from ancient writers as to the fact that such was the case.

Now the severe winters which Ovid[a] describes at the place of his banishment in the Crimea, where he represents the Danube covered over with ice solid enough to be crossed by horses, the very sea frozen so as to be walked upon, the plains destitute of verdure and wood, the apple-trees bearing no fruit, the inhabitants muffled up to their very throats in skins, seem to point to a climate more rigorous than what prevails at present in that country.

Similar is the picture given by Virgil in his Georgics of the climate on the Danube,—the wine frozen and cleft with an axe, the cattle sheltered during winter in houses, or else smothered in the snow.

I may be permitted, perhaps, to quote some lines from Sewell's translation of the Georgics in proof of this:—

> "But not where Scythia's hordes,
> And wave Mæotian, and that turbid flood,
> Ister, in eddies rolling golden sands;
> And where far outstretched Rhodope returns
> Towards the meridian pole; there keep they herds
> Preserved in folds, nor anywhere appear
> Or herbs in field, or foliage upon tree;
> But shapeless beneath snow-drifts, and deep ice,
> The earth lies all around, and rises high
> E'en to seven ells.   .   .   .   .   .

[a] See especially De Ponto, lib. iv. eleg. 7 and 9.

">   . . . . Crusts of ice
All of a sudden in the running stream
Shoot into masses, and the wave now bears
Steel-plated wheels upon its back; that wave
Erst broad-bowed boats, now welcoming the wains.
And brazen vessels oft asunder split,
And robes freeze stiff while worn, and liquid wines
They cleave with axes. And whole lakes have turned
To solid ice, and grim on beards unkempt
The icicle has hardened. All the while
Unceasingly through heaven entire it snows.
Perish the herds—thus stand all wrapt in showers
Of sleet huge shapes of oxen, and, in throng
Close gathered, harts beneath the novel weight
To numbness freeze, and scarce with antler tips
Above it peer [b]."

I quote this description, not of course in the same sense as I should have been justified in doing the report of a prose writer, in proof of these very phenomena having actually been witnessed on the Danube at the time to which Virgil

[b] "At non, qua Scythiæ gentes Mæotiaque unda,
Turbidus et torquens flaventes Hister arenas,
Quaque redit medium Rhodope porrecta sub axem.
Illic clausa tenent stabulis armenta; neque ullæ
Aut herbæ campo apparent aut arbore frondes;
Sed jacet aggeribus niveis informis et alto
Terra gelu late, septemque assurgit in ulnas:
Semper hiems, semper spirantes frigora Cauri.
Tum Sol pallentes haud unquam discutit umbras:
Nec quum invectus equis altum petit æthera; nec quum
Præcipitem Oceani rubro lavit æquore currum.
Concrescunt subitæ currenti in flumine crustæ,
Undaque jam tergo ferratos sustinet orbes,
Puppibus illa prius, patulis nunc hospita plaustris.
Æraque dissiliunt vulgo, vestesque rigescunt
Indutæ, cæduntque securibus humida vina,
Et totæ solidam in glaciem vertere lacunæ,
Stiriaque impexis induruit horrida barbis.
Interea toto non secius aëre ningit;
Intereunt pecudes, stant circumfusa pruinis
Corpora magna boum, confertoque agmine cervi
Torpent mole nova, et summis vix cornibus exstant."
(Line 349—370.)

refers, but as a proof, that countries in that latitude were then visited with such severe winters, as to lead the poet to single them out, as affording examples of what he pictured to himself to be the characteristics of a rigorous and almost arctic climate.

Other writers, too, speak in the same terms of the cold experienced in Gaul and Britain, and even the accurate Strabo states, that the Fig and Olive would not grow, and that Vines would not ripen their grapes, north of the Cevennes.

The freezing of the Tiber, as Hume remarks, seems not to have been an uncommon event in ancient days, whereas it is at present unusual for the snow to lie two days at a time in the streets of Rome[c].

Descending to modern days, we may perhaps obtain evidence of a greater rigour in the climate of these northern latitudes than is perceived at present.

Arago, in the *Annuaire* for 1828, has enumerated some of the most severe winters on record in France and England.

It appears that the Seine was frozen over in

1740, and the thermometer went down to			6.8	Fahr.
1744	,,	,,	14.0	,,
1760	,,	,,	15.8	,,
1767	,,	,,	3.2	,,
1776	,,	,,	10.4	,,
1788	,,	,,	8.6	,,

The greatest cold observed was in 1795, when on the 25th of January the thermometer sank to —10° below zero.

The next, on the 13th of January, 1709, when it fell to —9° 5′.

In England the winter of 1708-9 was, if not more severe, at least of longer duration, than we have ever since experienced; and during the course of the last century, several seasons have occurred of sufficient severity to cause the Thames to be frozen over.

It is true, that during the late severe winter of 1860-61,

[c] See Hume, Populousness of Ancient Nations, Essays, vol. i. p. 477.

more damage seems to have been done to the tender shrubs and evergreens than had been recorded on previous occasions.

These fatal effects, however, appear to have been owing to the occurrence of a great cold succeeding a moist autumn; for if we look at the extreme of temperature that occurred, we shall find it equalled in some preceding years.

Indeed, out of ninety years, from 1771 to 1861, recorded in Glaisher's Tables, no less than eighteen had a lower mean temperature than the one which we have lately experienced[d].

[d] The following is a list of some of the lowest mean temperatures recorded in the Tables alluded to, none being noticed excepting those below 35°; that of

		DEG. MIN.
1860-61 (mean of Dec., Jan., and Feb.) being		37.3
1837-8 ,, ,, ,,		34.3
1813-14 ,, ,, ,,		32.5
1799-1800 ,, ,, ,,		34.7
1798-9 ,, ,, ,,		34.4
1796-7 ,, ,, ,,		33.8
1794-5 ,, ,, ,,		31.6
1788-9 ,, ,, ,,		34.1
1784-5 ,, ,, ,,		32.5
1783-4 ,, ,, ,,		32.0
1779-80 ,, ,, ,,		34.7

If we take the mean winter temperatures for each period of ten years since 1771, we find that during—

				DEG. MIN.
1st decennial period, viz. from 1771 to 1780, Greenwich				36.94
2nd ,, ,, 1780 ,, 1791 ,,				36.52
3rd ,, ,, 1791 ,, 1800 ,,				36.8
4th ,, ,, 1801 ,, 1810 ,,				38.0
5th ,, ,, 1811 ,, 1820 ,,				37.3
6th ,, ,, 1821 ,, 1830 ,,				37.8
7th ,, ,, 1831 ,, 1840 ,,				38.2
8th ,, ,, 1841 ,, 1850 ,,				38.7
9th ,, ,, 1851 ,, 1860 Oxford				38.98

Curve of winter temperatures during the nine decennial periods.

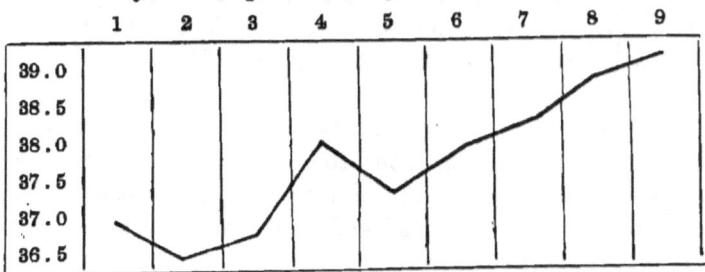

The mean temperature of London has been stated in p. 18, on the authority

The inference, therefore, I should be inclined to draw, would, upon the whole, be rather favourable to an improvement in the temperature of this country, than the reverse, notwithstanding the experience we have lately had of two winters of more than average severity.

The former existence of vineyards, indeed, in this country, if it be an admitted fact, might lead to the opposite conclusion; but these must at any rate have been confined to the neighbourhood of the rich monasteries, or the domains of the larger proprietors, where the climate may have been already mitigated by culture.

Even now, it would appear, from the statements of Boussingault, that the mean temperature of Hampshire "might be high enough to admit of vine culture," as well, perhaps, as the northern parts of Germany, where it is carried on.

At any rate, some amelioration in the climate of North America is stated on good authority to have taken place, since the country has been brought under tillage, and the forests reduced in extent.

It must, however, be admitted, that in some northern countries, such as Iceland and Greenland, a very marked deterioration of climate has taken place within historical times.

Sir H. Holland, in his "Dissertation on the History and Literature of Iceland," prefixed to Sir G. Mackenzie's Travels, observes, that it appears certain, that Corn was formerly grown in that island, and that the trees and shrubs at one time attained a much larger size, and were more numerous than at present,—both from the discovery of trunks in the morasses, and from the frequent mention made in the ancient writings of houses, and even ships, constructed of native timber. We read, too, of a feast in the western part of the island continuing for fourteen days, at which nine hundred persons were assembled; and of an entertainment given by two brothers in the northern province, where was an assemblage of four-

---

of Dove, at 50° 83'; but that of Greenwich is set down by Glaisher at 49°·6', or 1° 2' lower.

That of Oxford I have calculated as being, in a series of years, 0° 83' less than Greenwich, so that its mean temperature will be 48° 74'.

teen hundred guests. In Greenland, too, a colony from Norway was established in the ninth century, which continued to thrive up to the year 1408, when it was destroyed, through the neglect of the mother country, and the attacks of the Esquimaux. It had a bishop, twelve parishes, and two convents, till the year 1408, when all communication with East Greenland was cut off by a vast accession of ice; and from that time no intercourse with this country and the rest of the world has been possible.

These and other facts doubtless point to a deterioration of climate; but this is connected with some one of those great physical changes which have taken place from time to time in our globe, and which impress upon our minds the conviction of man's impotence, to counteract, in any great degree, the laws of nature, which he struggles to modify for his own advantage.

I may remind you, that the sinking by a few hundred feet of the great valley of the Mississippi, or the depression of a strip of the continent near the Isthmus of Darien, might divert the course of the Gulf-stream, and thus do infinitely more than all human efforts could hope to achieve towards modifying the climate of Western Europe.

"We should thus," as Hartwig remarks (p. 50), "not only lose the benefit of its warm current, but cold polar streams, descending further to the south, would take its place, and be ultimately driven by the westerly winds against our coasts.

"Our climate would then resemble that of Newfoundland, and our ports be blocked up during many months by enormous masses of ice."

"Under these altered circumstances, England would no longer be the grand emporium of trade and industry, and would finally dwindle down from her imperial station to an insignificant dependency of some other country more favoured by nature."

It is, however, a fact of deep significance, and one calculated to lead to solemn reflections, that whilst our utmost efforts to ameliorate our condition might be counteracted by any one of those great physical changes, which geology informs us have happened, and which, therefore, might occur again, Man, by the indulgence of his own malignant passions, produces often ten times as much mischief, as the most tremendous visitations of Providence, that we have ever witnessed, are known to have occasioned.

That the former of these propositions is based on truth will be admitted, when we reflect, that such effects as I have contemplated are by no means beyond the range of those powers, which, at the bidding, and under the control, of the great Author of the Universe, are working constantly beneath our feet.

The fact that during a period, geologically speaking, modern, the whole of the northern hemisphere, far below the latitude in which we reside, was covered with ice, leads to the inference, that owing to some of those great changes in the distribution of sea and land which we have abundant evidence of in other parts of the globe, the Gulf-stream, if indeed it had existed, was diverted from its present course; so that the very condition of things which I have contemplated must then have been realized.

But when we enquire into the character of the animated creation that had then been called into existence, we find that man was not then in being, and that the animals and plants of that period were such, as might be suited to the conditions of the climate which then prevailed over the northern hemisphere.

And with respect to the latter part of the proposition—setting aside as foreign to the present subject that great physical catastrophe of which Scripture speaks, and which, whatever might be its nature and extent, does not affect our argument, as we are told by the same Authority as that which relates it, that it is never to recur—we may remark, that a general review of history and tradition fully justifies us in recognising the greater intensity of those evils which man brings upon himself, as compared with those which are attributable to natural causes.

I am aware, indeed, of the fearful ravages that have been caused by pestilence in certain ages and countries, but this is a calamity in producing which *moral* as well as *physical* causes co-operate; and in order to estimate fairly the extent of the evil proceeding from the latter, it would be necessary in the first place to eliminate all that was due, to the filthy and vicious habits of society—to the indolence, the improvidence, the selfishness of man.

It was, indeed, from feeling how much natural evils are aggravated by human misconduct, that old Homer puts into the mouth of Jupiter the complaint, that man is always laying his misfortunes at the door of the Immortals, whereas he by his own folly is continually bringing upon himself calamities beyond what the Fates had decreed:—

> "Perverse mankind! whose wills, created free,
> Charge all their woes on absolute decree,
> All to the dooming Gods their guilt translate,
> And follies are miscalled the crimes of fate[e]."

But if we take the case of a physical catastrophe, over which man has no control at all, such, for instance, as an earthquake, it may be remarked, that one of the severest on record, namely, that which desolated the Neapolitan provinces in 1858, swept away only about 20,000 persons in all the different places which lay in its path, whereas the present war in America, even if it were brought to a speedy issue, will have cost the lives of half a million of the human race.

Well might David say, "Let us fall into the hands of the Lord, for His mercies are great, but let us not fall into the hands of men."

In considering these facts, we cannot but be impressed with the truth, that in this, as in other cases, it is by taking advantage of the arrangements of nature, rather than by struggling against them, that we can hope to advance in civilization, and to ameliorate our comforts and condition.

Let us therefore proceed to consider whether, if the cultivator is unable to do much towards accommodating the

---

[e] Ὦ πόποι, οἷον δή νυ θεοὺς βροτοὶ αἰτιόωνται.
Ἐξ ἡμέων γάρ φασι κάκ' ἔμμεναι· οἱ δὲ καὶ αὐτοὶ
Σφῇσιν ἀτασθαλίῃσιν ὑπὲρ μόρον ἄλγε' ἔχουσιν.
Odyss. I. 30—32.

climate to the vegetation, he may not effect something in the way of adapting the plants themselves to the nature of the climate.

To this latter branch of the enquiry belongs the question as to the power of man to acclimatize plants, which has long been a matter of dispute.

If the analogy between plants and animals be allowed to enter into the consideration, there would seem to be an antecedent probability, that such a change in the constitution of a plant, as should adapt it to a climate colder or warmer than its native one, might gradually be brought about.

The human race, it may be said, as well as many of the domestic animals which accompany him in his various migrations, prove themselves capable of subsisting in every climate, not from natural insensibility to differences of temperature, &c., for an individual brought up in one climate soon feels the injurious influence of removal into another, but from the race gradually accommodating itself to the new circumstances in which it is placed.

If then, it may be said, we admit that man is descended from a single pair, it must be assumed, that a process of acclimatization has been going on, by which the Negro on the one hand, and the Esquimaux on the other, have been gradually adapted to those excesses of heat or of cold, neither of which the inhabitant of a temperate region is calculated to support.

But the analogy fails in one important particular.

An animal, like man, possessed of a certain amount of intelligence, is capable of modifying to a great extent the effects of the climate in which he is placed.

The Negro reposes during the extreme mid-day heat, and by the influence of that insensible perspiration which takes place in all warm-blooded animals, keeps his body down to the normal point, even though the external temperature should exceed it.

The Esquimaux, on the other hand, by the aid of ample garments when out of doors, and by the joint effect of close apartments and artificial heat when within, as well as by taking advantage of the more equable temperature of the earth by burrowing under ground, contrives to maintain an invariable warmth within his own person during the most rigorous winter.

If, as is the case with the Fuegeans, and some other miserable tribes of savages, life is supported without these external appliances,—and if it be true, that in the former inhospitable region the native is almost destitute of clothing, and lives so unprotected from the weather, that he imprints the form of his naked person upon the snow on which he stretches during the night,—the stunted frames of these creatures, and their feeble and unhealthy physical organization, prove that there are limits beyond which no amount of training or discipline, though it be carried over many generations, can enable even man, the most flexible in his organization perhaps of all animals, to proceed.

But passing over these analogical arguments, let us examine how far facts support the idea, that plants are capable of being acclimatised.

It must be confessed, in the first place, that there is little or no evidence, that plants produced from the seeds of a tropical species, if sown in a region colder than that to which they are indigenous, acquire in consequence a different temperament, or a lesser degree of susceptibility to cold, than what is the attribute of the species in general.

Mr. Darwin indeed remarks, that the Pines and Rhododendrons which had been sown from seed collected by Dr. Hooker from trees growing at different heights on the Himalayas, were found to possess different constitutional powers of resisting cold; that Mr. Thwaites had observed similar facts in Ceylon; and Watson in England.

These facts, however, may perhaps admit of a different explanation; for that mere habit should have produced such a change of constitution as is here supposed, seems refuted by the case of the Jerusalem Artichoke, which Mr. Darwin with his usual candour adduces, a plant of which, as it has not been propagated by seed, new varieties have never been produced, and which is as tender now as it was on its first introduction into our country.

Notwithstanding, therefore, the contrary opinion of Darwin,

I am disposed to regard as chimerical the suggestion which has been sometimes put forth, namely, that the productions of a hot country might perchance be gradually naturalized in a cold one, by introducing them by successive steps into countries intermediate in point of temperature, between that of their native country and the one in which it is wished to establish them. I cannot, for instance, bring myself to believe that the Date-palm, taken from the neighbourhood of Nice, where it may be supposed to have already acquired a certain degree of hardihood beyond that which it possesses in Algeria, might be made to grow on the western coasts of France, as at Bordeaux; that from thence it might be transported to Jersey, and thus in the course of a certain number of generations might establish itself at Penzance or at Falmouth.

There are, I am aware, both in London and Paris, Acclimatization Societies, founded with a view to the introduction both of exotic animals and plants; and the latter, which is under the immediate auspices of the Emperor of the French, possesses a garden which, if it has no other merit, is at least a great ornament to the Bois de Boulogne.

Nevertheless, I am not aware that either here or in England the efforts of these Societies have been yet crowned with success; and undoubtedly, when I visited in winter the Paris establishment, in which the attempt appears to be that of maintaining the objects of experiment at the lowest temperature compatible with their life, the plants therein cultivated appeared to have suffered in vigour and luxuriance, in proportion as the temperature fell short of that point which it ordinarily attains in the regions to which they were indigenous.

Of course, the ready answer to this objection would be, that sufficient time had not yet elapsed to produce a sensible influence upon their temperaments; but at least I am afraid it must be admitted, that no facts can be as yet brought forward, founded upon the experiments proceeding in these Acclimatization Societies, which are calculated to shake our faith in the *à priori* arguments tending to shew, that the idea of hardening a tender plant, by gradual exposure to a severer climate, is chimerical.

But I believe the cultivator may in some cases succeed in acclimatizing a plant in a colder region, by selecting for his purpose the robuster varieties of the species he wishes to introduce.

Thus the late variety of the Walnut flourishes in localities where the early ones are killed by the frost. Thus the early sorts of Vine bear fruit in climates, where, owing to the insufficient heat of summer, or the early frosts of winter, the plant in general fails.

In like manner, by selecting the tubers of those Potatoes which blossom first, and by repeating that selection several times in succession, we may at length obtain a variety, which arrives at maturity in less than three months, and which therefore, as it would seem, might be introduced into climates where the summer is too short to ripen the ordinary samples.

The importation of Olive trees obtained in the Crimea, which appear less sensible of cold than those of the south of France, might perhaps enable us to extend the cultivation of that fruit beyond its present geographical limits.

In accordance with the above views, the Archbishop of Dublin some years ago suggested to me a scheme of acclimatising plants, which seems in some measure an anticipation of the great principle of natural selection, since brought so prominently forward by Mr. Darwin.

That the variation in point of vigour which we perceive in certain individuals of a species, in the vegetable as well as in the animal kingdom, is constitutional, may be proved by the experiments which the Archbishop had himself tried of grafting an early Hawthorn upon a late one, and *vice versa*.

In these cases the scions kept to their times of flowering (which differed by about a fortnight from that of the stock) just as before they were grafted; thus shewing that the difference between them was congenital, and not dependent upon soil and situation.

Proceeding, then, upon the assumption, that amongst every hundred individuals of a given species some are naturally hardier than others, he proposed to sow a number of seeds of each in a climate considered somewhat too cold for the plant in general to be reared in it.

It follows, of course, that the larger portion will die, but a few of the hardier individuals may nevertheless survive.

Repeating this process by sowing seeds obtained from these latter varieties, in a climate somewhat colder than the first, it is conceived that a robuster race might be produced; inasmuch as the survivors would be the hardiest varieties of the hardiest parents, and therefore might be capable of existing in a climate very much colder than the one natural to the species in general.

It is indeed in this way alone that we can suppose the human race to have become acclimatised in regions hotter or colder than the one in which it was originally planted; not that individuals became altered in constitution by being transplanted into a new country—for of this notion our Indian experience shews us the fallacy,—but that those endued with the greatest power of resisting heat in the one case, and cold in the other, survived; whilst those to which the climate was less congenial, perished.

But although the above is the only sense in which the term *acclimatisation* can, properly speaking, be applied to the vegetable kingdom, yet other means of cultivating plants under unsuitable conditions of climate have been suggested by horticulturists.

I need not particularly allude to the methods of preserving tender plants from cold by placing them under glass, and subjecting them to artificial heat.

These expedients appear to be as old as the Roman Empire; and from allusions made to them by Martial in his Epigrams, it might be conjectured that glass as transparent as what we have at the present day was manufactured for the purpose.

In one passage, indeed, Martial gives a description seemingly applicable to a modern orchard-house, where he says:—

> " Your oranges and myrtles, with what cost
> You guard against the nipping wind and frost,

> The absent sun the constant stoves repair,
> Windows admit his beams without the air [f]."

But in another passage, where the poet celebrates the garden of his friend Entellus, he gives equal praise to the "gemma" or translucent stone, with which the vintage was protected, but not concealed, comparing it to female beauty shining through silken folds—a simile much more applicable to Lapis specularis or Talc, than to the glass of modern days:—

> "Here, lest the purple branch be scorched with frost,
> And Bacchus' gifts by cold devouring lost,
> Shut in the *glass* (*gemma*) the living vintage lies
> Securely clothed, yet naked to the eyes;
> Through finest lace so female graces beam,
> Pebbles are counted in the lucid stream.
> What will not nature yield to human skill
> When sterile winter shall be autumn still [g]?"

It is indeed very improbable that glass should have been employed by the Romans for their conservatories, when we consider how very scanty a use was made of it by them for other purposes where transparency was required,—their windows being composed of Talc, or Lapis specularis, and Nero himself selecting a kind of stone called *phengites* for the purpose of glazing the windows of one of his temples.

Plates of glass indeed, taken from the baths of Pompeii,

[f] " Pallida ne Cilicum timeant pomeria brumam
  Mordiat et tenerum fortior aura nemus
  Hybernis objecta notis specularia puros
  Admittunt soles, et sine fæce diem."

[g] The force of my remark will be felt more forcibly by consulting the original, which I subjoin:—

"Invida purpureos urat ne bruma racemos
  Et gelidum Bacchi munera frigus edat,
  Condita perspicua vivit vindemia *gemma*,
  Et tegitur felix, nec tamen uva latet
  Fœmineum lucet sic per bombycina corpus:
  Calculus in nitida sic numeratur aqua.
  Quid non ingenio voluit natura licere?
  Autumnum sterilis ferre jubetur hyems."

See also Seneca, Epist. 90; Columella, xi. 3. 51; and Pliny, xix. 23.

are preserved in the Museo Borbonico at Naples, but they are only semi-transparent, and therefore calculated to transmit no more than that dim sort of light which was required in a bath.

Very imperfect indeed, even so late as a century ago, were the expedients for preserving plants under glass; and it is only since the improvements in the art of manufacturing that article have in so wonderful a manner reduced its price and extended its use, that in our stoves or conservatories anything more has been aimed at, than to minister to the luxuries of the rich, or to aid in the introduction of rare and tender exotics.

Of late, however, attempts have been made to render glass houses really available for the cultivation of our commoner fruits, by erecting Orchard and Peach-houses, which with little or no artificial heat secure these productions from the contingency of an early frost.

And something also has been already done towards the still more important object of establishing Sanatoriums, by the construction of gardens under glass, where the invalid might find in his native land those opportunities of taking exercise without exposure to cold, which he is at present compelled to seek in a distant land.

But without resorting to these expedients for introducing an artificial climate into our own country, much may be done by the horticulturist, by duly attending to the principles upon which the distribution of heat is known to be regulated.

Thus in selecting the stations in which an exotic is to be planted, it will be better that the spot fixed upon should be shaded from the sun, than that it should be exposed to its full glare.

In the latter situation the plant is rapidly thawed by the morning sun after a cold night, and suffers in consequence materially.

It is also stimulated into early growth, and therefore is more apt to be damaged by the late frosts of spring, a cause which has probably interfered with the cultivation of the Vine in the United States, where in the spring a degree of vicissitude in the temperature from extreme heat to extreme cold takes place, such as we have little or no experience of in Europe.

Thus during a tour I made in the Western States of North America in 1837, I noticed the thermometer on the 9th of April, at Little Rock, in Arkansas, standing at 83°; on the 15th, in the same district, at Washita Springs, it fell in the night to the freezing point, but rose in the day at Little Rock again to 83°, sinking to the freezing point on the following night; again, on the 19th of April, it fell in the night to 37°; during the succeeding days it ranged betwixt 68° and 84° in the shade, but during the month of May frequently sank during the night to 40°, and even sometimes to a still lower point.

This alternation of almost tropical heat during the day with a temperature approaching to freezing by night, which often takes place in these parts of America, cannot but exercise an injurious influence upon plants like the Vine, full of juice, and easily excited into action by the genial warmth of the day.

Much also may be effected by accommodating a plant to the new circumstances under which it is placed.

Thus, for example, the injury caused by the winter's cold may be greatly mitigated, by taking care that the plant shall be in as dry a condition as possible, so that it may be free from those juices which by their freezing cause the principal mischief.

This is done by selecting a soil as perfectly drained as possible, and a situation somewhat elevated, the former preventing the plant from gorging itself with moisture in winter, the latter securing it from those early frosts of autumn, or the late ones of spring, which prove so injurious.

In the seventh volume of the "Horticultural Transactions," Mr. Street has published the details of a method, by which he succeeded in growing tender plants in East Lothian, referable chiefly to the principles above laid down.

In this memoir he points out, that for tender exotics a loose sandy soil on a declivity presents the most favourable conditions.

He considers, moreover, that cuttings are in general more hardy than seedlings, and that the seeds of annuals usually sown in hotbeds, produced plants of a hardier character, when sown in a warm situation in the open ground.

Another circumstance to be attended to in the treatment of plants introduced from warmer countries than our own, is not to over-tax their energies by allowing them to rear too large an amount of fruit.

Setting out with the principle, that the vigour of a plant is in a direct proportion *cæt. par.* to the warmth of the locality, and that in a cold climate there is the same tendency in it to put forth branches and fruit without the same capacity to mature them, it will be wise to adopt the plan recommended by a skilful cultivator of the Vine, who prunes at least seven-eighths of the shoots and branches that make their appearance upon the tree.

Another method, which has been strongly recommended by Dr. Lindley in the "Gardener's Chronicle," consists in pruning away several of the roots, by which means the quantity of sap drawn up is diminished, so that no more is carried into the stem and branches than the energy of the plant enables it to elaborate. Hence it does not waste itself in producing branches and leaves instead of fruit, and is less liable to freeze from not being filled with so large a quantity of aqueous juices. This is a good plan to pursue with fruit trees that are not good bearers.

With regard to the description of plants most capable of being introduced into other climates than their native ones, it may be observed, that annual and other herbaceous species are best fitted to be naturalized in a temperate region, because their existence is limited to a few months' duration, in which the temperature may approach that of their native country. And if it be said, that the heat, even during these months, rarely rises to the point which it attains in the tropics, it may be replied, that this will not be material, provided the continuance of the heat be long enough to compensate for its want of intensity.

I have already in a former Lecture alluded to the rule

laid down by Adanson, namely, that each plant requires a certain definite amount of heat to cause it to put forth flowers.

Thus, he says, the White Poplar comes into flower when 168° of heat have entered it since the 1st of January, the Lilac when 723°, the Vine when 1,770° have been communicated to it.

The method by which Adanson arrived at these results is open doubtless to objection; for, as Decandolle the elder pointed out, the 1st of January is an arbitrary point of departure from which to begin calculating the temperature received by the plant; it being probable, that the degree of warmth present in the autumn will influence the plant likewise, and that each species will not have arrived at an equal degree of development on the day which is taken as the starting-point. Nor ought we to leave out of the account the relative coldness of the nights, or the degree of brightness of the sun, during the period to which the plant is exposed to its influence.

Still, the principle laid down by Adanson seems to be so far true—inasmuch as each plant requires to absorb a certain amount of heat in order to arrive at its due development—inasmuch as this amount varies in each particular case,—and as it is immaterial, comparatively speaking, whether the requisite amount be obtained in a short time by an intense temperature, or in a longer time by a more moderate one.

Accordingly, in Russia, where the day in summer is both longer and hotter than with us, the Barley is sometimes fit for reaping within forty days from the period of sowing, whereas in our climate about 150 days are considered requisite [h].

Plants which die down every year, but possess tuberous or bulbous roots, which are vivacious, as is the case with the Lily tribe, become naturalized more easily from a warmer to a colder climate, than the reverse.

Thus the common and the sweet Potato, both natives of hot climates, are cultivated with more or less success over the greater part of Europe; whereas the Asparagus, a native

[h] See above, Lect. iii. p. 96.

of this and other temperate regions, when transplanted to India, speedily perishes.

Plants whose stems grow uninterruptedly throughout the year, such as the Banana, Crinum, &c., cannot be naturalized in a country subject to occasional frosts; for being filled with juice during the winter, they are liable to be destroyed by the rupture of the vessels occasioned by the congelation of their aqueous contents.

On the other hand, trees from the warmer regions of the globe, which have certain alternations of rest and activity in their circulation, may often be introduced into colder ones.

Thus we owe to Persia the Walnut and Peach; to Armenia the Apricot, and perhaps the Vine; to Asia Minor the Cherry and the Chesnut; to Syria the Fig, the Pomegranate, the Olive, and the Mulberry. Even the Paper Mulberry, Broussonetia papyrifera, brought from the Society Islands, thrives in sheltered parts of this country.

The greater tenderness of evergreen trees, where the vegetation goes on uninterruptedly throughout the year, may be estimated, by the necessity which exists for preserving the Orange and Lemon under shelter in all but the warmest parts of Europe.

Thus in Louisiana, where the summer temperature is tropical, and admits of the cultivation of the sugar-cane, the Orange and Lemon trees are frequently cut off by the cold of winter, as happened only a few years ago.

It is truly remarkable to meet with plants which, properly speaking, cannot be said to be naturalized in North America in latitude 30°, flourishing in the open air at Falmouth in latitude 50°.

I have seen in Mr. Fox's garden at Grove-hill, near that town, Orange and Lemon trees bearing fruit,—in one case planted out in the common border, and in others covered over only by a mat in seasons of peculiar severity.

I have seen in the same garden at least ten shrubs, usually regarded as stove plants in this country, flowering freely, and growing vigorously in the open air, with no other protection than a mat.

In the Abbey gardens, at Tresco, in the Scilly Islands, belonging to Mr. Smith, M.P. for Truro, a still nearer approach to the vegetation of southern climes is exhibited; and as I have been favoured with a list of plants grown in both these localities, I shall insert them at the close of this little volume[l], as likely to convey a more faithful idea of the remarkable mildness of either spot, than could be imparted by any more general description.

A yet more remarkable deviation from the ordinary laws of climate,—as not being limited, like the former, to a particular locality, but spreading over an entire region,—occurs in Terra del Fuego, a country situated in latitude 45°, of which the mean temperature, during the hottest month, is only 51°, and during the coldest 33°, as we learn from Captain King and Mr. Darwin; notorious for its stormy and inhospitable climate, and as it would appear, most unfavourable to the healthy development of man and the higher animals; and yet in this inhospitable tract Captain King[m] describes a luxuriant vegetation—large-stemmed trees of Fuchsia and Veronica, in England treated as tender plants, in full flower, within a short distance of a mountain covered for two-thirds down with snow, and with the temperature at 36°.

A little higher, on the island of Chiloe, in latitude 42°, Mr. Darwin represents the character of the Flora as almost tropical. Stately trees of many kinds, with smooth and highly coloured barks, were loaded with parasitical Orchidaceæ, large and elegant Ferns were abundant, and arborescent Grasses entwined the trees in one entangled mass, to the height of thirty or forty feet above the ground.

Yet, he says in another place, our fruit-trees rarely ever ripen their produce in this climate, and the inhabitants are frequently driven to cut down their corn before it is ripe, and to bring it into their houses to dry.

The above remarks on the naturalization of plants may possibly appear to have but little practical bearing upon the conduct of an English farm, since in a country, every portion of which, with the exception of a few mountainous tracts,

---

[l] See Appendix, No. II. [k] Geograph. Journ., 30 and 31.

enjoys a summer temperature high enough for the cultivation of all the kinds of Cerealia alluded to, the question of climate may be thought hardly fit to enter into the calculations of the farmer in the allotment of his crops. Nevertheless, even to him it may be useful to possess the data for estimating the influence which a summer, warmer or colder, wetter or drier than ordinary, has exerted upon the productions of his farm, so as not to be misled in his calculations as to the advantages or disadvantages of any novel plan of cultivation.

These data have in part been supplied by Mr. Lawes in an elaborate paper published in the eighth volume of the Royal Agricultural Society's Journal for 1848, in which he shews, that in 1844, 1845, and 1846, the difference in the amount of produce was in accordance with the general character of their respective seasons.

Thus it will be seen by turning to the table referred to in the note [n], that in 1844, when there were only 81 rainy days, and when the mean summer temperature was 57° 5', the farm produced 16 bushels of Corn to the acre, and the weight of the bushel was $60\frac{1}{2}$ lb., whilst the grain bore to the straw as high a ratio as about 82 to 100.

In 1846, when there were 93 rainy days, and when the mean temperature was as high as 59°, the yield was about 17 bushels per acre, and the weight of a bushel 68 lb.; but the proportion of grain to straw was lower, namely 76 to

[n] The following table indicates the effect of climate upon the quantity and quality of the produce of the unmanured piece of the experimental wheat-field, (during three seasons); the average results of the variously manured, &c., are also given:—

	1844	1845	1846
Corn per acre in bushels	16	23	17.25
Straw per acre in pounds	1120	2712	1513
Weight of Corn per bushel in pounds	$58\frac{1}{2}$	$56\frac{1}{2}$	$68\frac{1}{4}$
Per centage of Corn to straw, (straw 1,000)	821	534	797
Mean of all the plots.			
Weight of Corn per bushel in pounds	$60\frac{1}{2}$	$56\frac{1}{2}$	63
Per centage of Corn in straw, (straw 1,000)	868	490	765
Mean temperature	57° 5'	55° 2'	59° 1'
Rainy days in $30\frac{1}{4}$ weeks	81	110	93

100, shewing that the yield of Corn had been influenced by temperature, but that the quantity of straw had been increased by the amount of rain.

Lastly, in 1845, when the number of rainy days was greater, though the temperature had been lower than in either of the two other years,—the former being 110, the latter 55,—the yield was much greater, amounting to 23 bushels per acre, but the weight of the corn per bushel less, namely 56 lb.; and the increase in the produce of straw such, that the grain only bore to it the proportion of 49 to 100.

These figures appear to shew, that although the greater quantity of rain was favourable to the amount of grain, yet that it tended to increase in a still greater ratio that of straw; and that the higher the temperature of the year was, the heavier the grain would prove, so as to make up in quality for its deficiency in actual quantity.

Thus it would appear, that under the same treatment, the produce may vary in the proportion of 7 bushels per acre, according to the difference of season, which is equivalent to one-quarter of the normal produce; or, calculating this at 30 million quarters in a good year, may be as much as $7\frac{1}{2}$ million deficient.

The above statements may likewise assist him in selecting proper positions for any new and tender plants which he may be disposed to introduce; for when we recollect that all our cultivated plants are natives of warmer climates, it cannot be considered chimerical to imagine, that other articles of culture may hereafter be discovered suitable to this country. To descend to particulars.

The Maize, where it succeeds, yields a larger return than either Wheat or Barley, and it is probable that in some warm spots it might be cultivated even in England. But the success of the experiment would turn upon the selection of a spot, combining in the greatest degree those conditions which most favour the absorption of heat.

Thus, too, to take the instance of a plant of native growth, the Hop is commonly regarded as a very capricious kind of crop, for although it be a native of almost all parts of Europe, it admits of successful cultivation only in a few.

In England it is limited to the southern and western portions of the kingdom; from which circumstance it might be inferred, that something beyond the average summer temperature of our island was congenial to it.

The blight to which it is peculiarly liable seems connected with the want of sun rather than with deficiency of heat, so that solar radiation would seem to be one of the conditions essential to its healthy and vigorous constitution.

Hence it appears clear that considerations of climate ought to be taken into account by those who contemplate engaging in the precarious speculation of hop-growing.

Again, to take the case of a plant which, although not indigenous, has been from time immemorial in a certain sense naturalized in this country.

I allude to the Vine, which affords an example of a plant, in the culture of which attention to the characteristics of local situation would afford material aid.

Unpropitious as the climate of England is pronounced to be to the ripening of the grape, there can be little doubt that vines were formerly grown in some of the southern counties; and it seems therefore probable, that by selecting certain favoured spots enjoying a higher summer temperature than common, wine not inferior to what is made in some of the northern parts of the Continent might be produced from our own vineyards.

Boussingault, from observations made in Alsace, concludes, that the mean temperature of the period of the year during which the vine is producing its fruit, namely, from April to October, ought to be as much as $61°$. Now it so happens, that the summer temperature of Gosport is quoted as above $62°$, so that the vicinity of this spot at least would seem to come within the limits assigned for the successful cultivation of the Vine.

At any rate, until these points are more fully settled, it cannot be said that the consideration of climate is alien to an agriculturist placed in the southern portions of this country; whilst in the northern, many cases will occur, in which the cultivation of Wheat, Barley, or Oats, will probably be determined by the local circumstances, which tend to impart to the particular locality more or less than the average degree of temperature or of humidity.

Considerations of climate may also be of great importance to the horticulturist. Cold air being heavier than warm, the stratum next to the soil will, as a rule, be colder than the one above it.

Hence land at the bottom of a valley will be chilled by the descent of cold air more than that higher up, so that what are called sheltered places are often in spring and autumn the coldest.

The Dahlias, Potatoes, and Kidney-beans of the sheltered gardens in the valley of the Thames have often been killed by frosts, whose effects were unfelt in the low hills of Surrey and Middlesex.

Professor Daniell says, he has seen a thermometer on the same night stand 30° higher on a gentle eminence than in a valley below.

This, indeed, was remarkably illustrated during the severe winter of 1860-61, when it was found, that low-lying parts of Scotland experienced the cold in a greater degree than the more elevated districts.

For example, Braemar and other places in the upper parts of Aberdeenshire had the thermometer at 8° or 11° above zero, when in the lower parts of the same county it sank —6 degrees below it.

In Dumfriesshire the thermometer fell below zero, but at Wenlock Head Station, 1,330 feet above the sea's level, the lowest marking was 6° above zero.

The coldest spot in England during this frost was Nottingham, where the thermometer indicated —11° at 4 feet from the ground, and —13° upon the grass.

Hence the advantage of placing a garden upon a gentle slope, especially when there is a running stream at its foot, which, at the same time that it presented a surface not liable to refrigeration, would also prevent an injurious stagnation of the air.

Dr. Lindley also states, that a south-eastern exposure is by no means the most favourable position for plants in general; for, as he remarks, the advantage of receiving the early sunbeams is counterbalanced by the exposure to easterly winds, which are the coldest and driest of any.

Moreover, the sudden action of the sun's rays is very detrimental to vegetables, that have been frozen by the cold occasioned through the radiation of heat from their surfaces during the preceding night; which, according to the explanation offered by Dr. Lindley, causes the air contained within the tissue of the plant to be expelled by the contraction of the fibres which then takes place.

The expelled air is consequently forced into parts not intended to contain it, and is there expanded rapidly by the sudden warmth of the sun.

This increases the disturbance already produced in the minute vessels by its expulsion from the parts properly intended to contain it, and a rupture of the vessels will often in consequence supervene; whereas if the thaw had been gradual, the air would have had time to retreat from its new position, without producing any additional derangement of the tissues.

It is also possible, that leaves from which their natural air has been expelled in the act of freezing, may from that circumstance alone have their tissue too little protected from the force of the solar rays, which we know produces a specific stimulus of a powerful kind upon their organs[o].

Such, then, are the facts with respect to the connection of climate with agriculture, which I have endeavoured to place before you.

They are defective, indeed, in many particulars which might assist us in deducing practical inferences from them, such, namely, as the influence of a greater or less intensity of solar light, of exposure to humidity, and of radiation, upon the crops noticed.

What, for instance, would be the effect of bright clear nights alternating with warm sunny days, as compared with that of a cloudy atmosphere possessing the same average temperature? What of a climate highly charged with hu-

[o] Hort. Trans., New Series, vol. ii. p. 305.

midity, as compared with one which from its dryness promoted evaporation from the surface of the vegetable?

Considering, then, in how many ways vegetation is affected by the conditions of the atmosphere, and how much a climate may differ in its effects upon plants, not only from its containing more or less vesicular vapour or mist, but also according as it retains suspended in it more or less water, even though it be in a condition in which it in no degree affects its transparency, a farmer or a gardener would do well, if he were to take the trouble of registering daily the weather upon some uniform system, noting at the same time the quality and quantity of his crops during the same period.

By this proceeding he might hope to arrive at many interesting results, by which the observations of Lawes and others, to which I have already directed your attention, would be confirmed or corrected.

For this purpose the temperature should be noticed at certain fixed times of the day, in the manner pointed out in my first Lecture.

The amount of radiation should also be observed, and the effects of any remarkable degree of intensity in the solar power upon particular plants should be carefully noted.

Professor Daniell, in one of his meteorological essays, has set us a good example as to the mode of conducting such a register; and has pointed out the probability of obtaining from records of this kind some insight into the causes of the various blights which affect the productions of the soil, whether it be the mildew and smut of Wheat, the fungus which attacks the Vine and the Potato, or the so-called fly of the Turnip and the Hop.

Thus he has traced a connexion between the relative force of the sun's radiation in the years 1821 and 1822, and contrasted the mildewed condition of the crops of Corn in the former or more cloudy season, with the healthy appearance they presented in the latter or brighter summer above mentioned.

The relative dryness of the atmosphere should likewise be tested by Daniell's hygrometer, or the more simple contrivance of Mason, called the wet-bulb, which I have already explained.

And lastly, the amount of rain and the direction and force of the wind should each day be entered in appropriate and distinct tables.

Having then once ascertained, what in point of fact are the circumstances of climate which act prejudicially upon the particular crops which he grows, the agriculturist in any experiments he may be led to institute would be able to discriminate, how much of any observed difference in the amount of produce was referable to the season, and how much to any new modes of culture which might have been introduced. He might also learn, how to lessen or remove the disadvantages of the climate under which his operations were carried on, by a judicious choice of soil and situation.

Is the temperature of the place insufficient for any of the crops he cultivates, he will recollect that wet tenacious soils are of all others the most difficult to heat and to drain.

Is it excessive, he will know that light sandy soils part with their moisture most readily, and absorb the greatest proportion from the sum of heat communicated by the sun.

A register of the weather, conducted upon the plan recommended, would also be of considerable utility in a sanitary point of view, both as affording a clue to an understanding of the causes which impart to the climate of the place in which it is carried on its peculiar character, and also as a guide to direct us in the search after other spots, likely to possess the qualities which we prize in the one under our consideration.

For it is not sufficient that we should be acquainted, in a vague general way, that the mean temperature of a particular spot is higher, or that the cold of winter is less severe than that of other contiguous places, since it is a combination of various meteorological conditions, more or less propitious to health, which stamps upon the climate its peculiar and characteristic physiognomy. Amongst these conditions — for I will not seek to enumerate them all—are, first, a temperature not subject to sudden vicissitudes, for the most part genial, and seldom, or never, sinking to any extreme degree of cold; secondly, an atmosphere sufficiently charged with moisture, in a state of perfectly transparent vapour, as to exert that soft and soothing influence over the lungs and skin, which, according to the views put forth by Professor Tyndall, will arise from the impediment it offers to the escape of heat from our persons; and yet exempt from fog and mists, so as to convey no such sensation of rawness, as we are apt to experience in other places, at times when bronchial diseases and rheumatic affections prevail; and thirdly, transmitting such a supply of sunshine, even during the winter months, as may be sufficient to exercise a cheering influence over the nerves and spirits.

Thus, although I have every reason to speak well of the climate of Torquay, as having carried me over two winters, without my being a sufferer from those bronchial annoyances, which it has been my lot to encounter in other parts of England during the worst portion of the year; it would be hasty in me, were I to attribute its superiority simply to its relative temperature, seeing that, during the present winter, the differences in this respect existing between this and other spots, even in the interior of the island, have been upon the whole not very considerable, as may be collected from a register containing the maximum and minimum temperature of every twenty-four hours during the last six weeks [p] at

[p] In lieu of the Register exhibited at the Lecture, I will now present one including the months of December, 1862, and of January, February, and March, 1863, and exhibiting the mean temperatures at three different places, viz., at Oxford, Bath, and Torquay, from which it will appear, that except during December, the difference in favour of the latter over the two former was not very remarkable.

The temperatures for Oxford are given on the authority of Mr. Main, of the Radcliffe Observatory; those for Bath by the Rev. L. Jenyns, from observations taken by him daily at 9 A.M. at Darlington-place; those for Torquay are deduced from the mean maximum and minimum of each month at Woodfield, supplied to me by Mr. E. Vivian. The latter are rather higher than those taken by myself at Torre during a part of this period. The following are the results:—

	OXFORD.	BATH.	TORQUAY.	
	DEG. MIN.	DEG. MIN.	DEG. MIN.	DEG. MIN.
December	43.9	43.8	47.2	mean max. 51.0 / „ min. 43.4
January	41.2	41.2	43.30	mean max. 46.4 / „ min. 40.2
February	42.6	41.2	44.35	mean max. 47.7 / „ min. 41.0
March	43.9	43.0	45.40	mean max. 50.2 / „ min. 40.6

Oxford and at Torquay. Nor, on the other hand, could I venture at any season to ascribe to this place the mildness of climate which belongs to Falmouth or Penzance, where, as I have already stated, tender, and even sub-tropical plants, of which this part of Devonshire cannot boast, flourish in the open air.

Nor assuredly will the brilliancy of the winter sun, cheering as it may be in comparison with other parts of England, compete with that which greets us in Italy, and in the south of France, where days are often met with, even in December and January, which we should be only too glad to transfer to the most genial month in the most highly favoured locality throughout Great Britain.

It would not become a mere bird of passage like myself, to speak with any confidence with respect to the peculiarities of climate belonging to a place in which he has taken up his temporary abode; and I shall therefore do best by referring you to the observations of residents, carried on during a considerable period of time, such as those for which we are indebted to the diligence and perseverance of your President, Mr. Vivian, who, for the last twenty years, has kept an accurate register of the weather at Torquay. From his published reports it would appear, that Torquay is colder than London in the summer, but that it is no less than 5° warmer in winter, at which season, indeed, it does not fall short of Rome by more than 2° 8′. With regard to rain, it has the advantage over many parts of the United Kingdom, so far as concerns the number of wet days it experiences, which bear the proportion to those of London and its vicinity of 132 to 178; whilst in relation to the quantity of rain that falls, it would appear that the number of inches annually collected may be reckoned at 28, being 4 inches indeed more than the rain-fall in London, and also exceeding in quantity that of the eastern coasts of England; but, at the same time, only 2 inches more than Oxford, which is calculated at about 26 inches, and 16 less than Penzance, which is stated as being 44.

The dryness of the atmosphere here is said to be greater than at Clifton; and, moreover, Torquay enjoys an enviable exemption from violent thunderstorms, the absence of which is an index of a tranquil and equable condition of the atmosphere, and, perhaps, also points to the existence of other meteoric conditions favourable to the temperament of invalids.

From these data I should infer, that the excellence of the climate of Torquay depends in part upon the absence of extremes of temperature; upon the general clearness and medium quality of its atmospheric condition, equally removed from excessive dryness and excessive humidity; from the less frequent occurrence and shorter duration of fogs, and particularly from a comparative exemption from those emanating from the land; and, moreover, from the presence of a larger amount of solar radiation than is to be met with in winter over most other parts of our island.

But I should also be inclined to attribute the advantages of Torquay in part to the nature of the rock formations upon which the town and its suburbs are erected, and to the general disposition of the houses upon an inclined plane; from which arises, not merely the dryness of the roads and footpaths, even soon after a fall of rain, but also a great facility for good drainage, and, in consequence, a general absence of morbific exhalations.

It is remarkable, that Birmingham, notwithstanding its crowded, and in some respects ill-conditioned population with respect to cleanliness and general habits of living, escaped the visitation of cholera on two occasions when it was raging all around it, owing, as I conceive, to the houses being in a great degree built upon a slope, and to the subsoil possessing the quality of absorbing moisture of all kinds, in consequence of being composed of porous red sandstone.

Something, too, must be attributed to the character of the dwellings themselves which compose the town and its environs, of which the better kind so often stand in the midst of gardens and pleasure-grounds, or else in detached blocks, with a free circulation of air around them; whilst even those of the poorer classes, with some exceptions, are scattered over a larger space of ground than in most other places, and therefore present fewer of those foci of contagion which are apt to be engendered by large communities when crowded within a small compass.

There is also a great advantage connected with Torquay, in the variety of situations which it presents, so differing one from the other in point of exposure, height, and consequent warmth, that the invalid can seldom be at a loss to find somewhere or other a climate suited to his particular constitution.

And as the temperature of different spots even in Torquay itself differs considerably, it may be useful to have observations taken simultaneously in different parts of the town, a task, which I am happy to find has been undertaken by Dr. Becker, a young physician recently established in this place.

Possibly, as I suggested in a former Lecture, the large proportion of ozone which has been noted in the air of this place may be connected with its purity, since there is reason to suppose, that what is detectable in the atmosphere represents the excess remaining, after the removal by oxidation of those exhalations which arise from animal or vegetable impurities.

But it is time for me to bring to a close remarks, which could only come with any authority from a person long resident in the place to which they relate, and therefore, with my best thanks to those who have honoured me with their attendance upon this and on the three preceding occasions, I will conclude, by expressing my satisfaction at finding, that I had not over-estimated the interest, which a Community, mainly drawn together to this place by the attractions of its climate, was likely to feel in a discussion, relating to the general laws which affect the weather, and to the influence which the latter exerts upon the vegetable and animal economy.

APPENDIX I.

(See p. 20.)

Effects of the Winter of 1860-61 on the Plants at the Botanic Garden, Oxford.

I EXTRACT from the "Gardener's Chronicle" for Dec. 7, 1861, the following particulars of a report communicated by myself respecting this remarkable winter; and have appended to it under the initial C a statement of the effects which it produced at Cornbury Park, Oxfordshire, as reported to me by Lord Churchill, the proprietor:—

It will be understood, that all the plants enumerated refer to the Botanic Garden, the addition of the letter C. being intended merely to express, that they met with the same fate at Cornbury.

Cupressus, sempervirens C, torulosa C [a], macrocarpa C, Tournefortii C [b], funebris, and Goveniana C, destroyed [c].

Deodars, three fine specimens killed [d].

Cedrus Libani C, and Atlantica, injured, but recovered.

Juniperuses, all the exotic species killed, except chinensis [e], virginiana, and squamata.

Taxodium sempervirens [f], injured, but recovering.

Pinus Llaveana C, insignis, macrocarpa, and Pinea, killed [g].

Cryptomeria japonica killed [h].

Quercus Suber killed, Ilex severely injured [i].

Lucombe oak severely injured, but has since recovered.

Common Laurel C, Portugal Laurel C, Laurustinus C, and Bay Laurel C, all cut to the ground.

Rosa Banksii destroyed, and severe havoc made amongst the standard Roses in general.

Phillyreas C, much injured, but recovering.

Photinias C, killed.

Arbutus Unedo C [k], killed; Andrachne, a fine and old tree, so nearly destroyed as to be cut down [l].

Fig-trees killed down to the roots.

Magnolia grandiflora C, killed down, but since putting out from the stem.

Cercis siliquastrum [m], a tree of thirty years' growth, killed.

[a] All but one. [b] Much damaged. [c] Also Cupressus thurifera and Uhdiana at C. [d] Much damaged at C. [e] At C escaped. [f] At C escaped. [g] At C two plants escaped, two killed. [h] Also P. Montezumæ and muricata at C, and P. Brutea, much damaged. [i] At C some unhurt. [k] Some escaped with damage at C. [l] Arbutus procera at C much damaged. [m] At C escaped.

Ligustrum japonicum, lucidum C, killed.
Pittosporum Tobira, do.
Euonymus japonicus C, do.
Garrya elliptica, standard at C killed, against a wall at C escaped.
Bupleurum fruticosum [n], killed.
Ceanothus azureus, do.; pallidus, do.; papillosus, do.; rigidus, do.
Buxus balearica [o], do.
Ribes echinatum, do.
Morus rubra, do.; multicaulis, do.
Escallonia macrantha, do.
Arundinaria falcata, severely injured, but since recovered.
Rhamnus Alaternus, do.
Hydrangea quercifolia, do.
Leycesteria formosa, do.
Ulex europæus C, killed.
Yuccas, old plants killed to the ground [p].

APPENDIX II.

(See p. 129.)

INCLUDING a list of the tender plants growing in Mr. Smith's garden at Tresco, in the Scilly Islands, and at Mr. Robert Fox's, either at Grove Hill, or at Pengellert, near Falmouth, the former being distinguished by the letter S appended to the name of the plant, the two latter by the letter F.

No plants are mentioned in this list, which are not too tender to be grown in the open air at the Botanic Garden in Oxford.

Where the name given is unknown to me a ? is added.

Abutilon Bedfordii, S; vitifolium, S; venosum, S.
Acacia affinis, F; lophantha, S and F; melanoxylon, S; nigricans, F.
Adenandra uniflora, S.
Agapanthus, S.
Aloe, several species, S.
Aralia Sieboldtii, S; quinquefolia, S; trifoliata, S; crassifolia, S; papyrifera, S.
Araucaria Bidwillii, S; Braziliensis, S; excelsa, S.[q]

Arundinaria falcata, S.
Aster argophyllus, S and F.

Banksia ericifolia, F.
Beaufortia decussata, F.
Bignonia jasminifolia, F; semperflorens, F.
Brachyglottis repandus, S.
Brugmansia suaveolens, F.

Calothamnus lineiformis, S; quadrifida, F.

---

[n] At C much damaged.   [o] Escaped at C.   [p] At C. escaped.
[q] Lately blown down by a violent gale.

Calendula japonica, ? S.
Camellia japonica, F.
Cassia corymbosa, S and F.
Cantua dependens, S.
Chamærops humilis, S and F; excelsa, S.
Chrysanthemum trifurcatum, S; frutescens, S.
Cineraria arborea, (and petasites?) S; large-leaved, F.
Citrus communis, F; medica, F.
Clethra japonica, S; arborea, F.
Clianthus puniceus, S. and F.
Cluytia pulchella, F.
Convolvulus Cneorum, S.
Cordyline rubra, S.
Coronilla glauca, F.
Corræa alba, S; ferruginea, S; pulchella, F; rubra, S; speciosa, S, F.
Cotyledon ovata, F.
Crassula coccinea, S; jasminea, S; lactea, S; orbicularis, S; portulacea, S.
Cupressus funebris, S.
Cytisus algavensis, S.

Dactylis cæspitosa, S.
Dammara australis, S.
Diosma cordata, F.
Dracæna australis, S; Draco, S; fragrans, F; indivisa, S, has flowered twice, one plant about twelve feet high.
Drimia indivisifolia, ? F.

Edwardsia grandiflora, F; macrophylla, S; microphylla, S.
Elæagnus reflexus, S.
Escallonia floribunda, S; organensis, S.
Eryobrotia japonica, F.
Eugenia australis, S and F; Ugni, S and F.
Erica colorans, F; gracilis, F; lævis, F.

Erythrina Crista-galli, F.
Eurybia ilicifolia, S; purpurea, S.
Eucalyptus coccinea, S; robusta, S; saligna, S.
Eutaxia myrtifolia, F.

Fitzroya patagonica.
Fuchsia, 14 species or varieties, S.

Genista canariensis, S.
Gnidia simplex, F.
Grevillia rosmarini folia, F.
Griselinia littoralis, S; lucida, S.
Grewia occidentalis, F.
Gunnera scabra, S.

Habrothamnus elegans, S; Hezelii, ? F.
Hedychium flavum, S.
Heimia salicifolia, F.
Hibbertia, S.
Hydrangea involucrata, S.
Hypericum chinense, F.

Jasminum revolutum, F.

Kennedya macrophylla, F.

Lardizabula bifurcata, S.
Lasiopetelum ferrugineum, F.
Laurus Camphora, S and F.
Leptospermum australe, S; ambiguum, F.
Lomatia aromatica, S.
Lophospermum erubescens, F.

Mesembryanthemum, 10 species at F, and 56 other species besides the above in S.
Metrosideros robusta, S.
Myrsine undulata, S.
Myrtus trinervis, S.

Nerium Oleander, F.

Olea europæa, S.

Œdera prolifera, F.
Opuntia cylindrica, S; exuviata, S.
Oxalis floribunda, F; lobata, S.
Oxylobium calycinum, S.

Passiflora edulis, F.
Pelargoniums, various, in great luxuriance, S and F.
Phygelius capensis, S.
Philadelphus mexicanus, S.
Phormium tenax, S.
Pitcairnia chilensis, S.
Pittosporum Toberi, S and F; tenuifolium, S; undulatum, F.
Plumbago capensis, S and F.
Podocarpus asplenifolius, S; chilensis, S; ferrugineus, S.
Polygala grandiflora, F; latifolia, F; myrtifolia, F; spinosa, F.
Polygonum Sieboldtii, S.
Prostanthera Lasanthus, S.
Psoralea pinnata, F.

Quercus glaber, S.

Sedum arboreum, S; canariense, S.
Solanum ferox, S.
Sparmannia africana, S.
Statice Dicksoniana, S; lavendulacea,? S; Rheinwardtii, S.
Salvia chamædryoides, F; involucrata, F.
Sutherlandia frutescens, F.
Swammerdammia antennaria, S.

Tasmannia aromatica, S.
Tecoma capensis, F.
Thunbergia coccinea, F.
Tetranthera japonica, S.

Veronica Andersonii, S; decussata, S; Lindleyana, S; salicifolia, S; speciosa, S; variegata, S.
Viburnum awafuhi, or japonica, S.
Vitex littoralis, S.

Woodwardia angustifolia, F.

DR. DAUBENY considers it due to the Subscribers, and to others who may possess his Lectures on Climate, to communicate to them the corrections and additions which will be introduced into any future impression of his Work that may be called for; especially as he is indebted for several of them to the kind communications of friends, whose names alone will be a sufficient guarantee for the authenticity of the statements given. The leaf now sent may be inserted in the place of that containing the List of Errata.

BOTANIC GARDEN, OXFORD,
July 4, 1863.

CORRECTIONS AND ADDITIONS.

Page 58, sixteenth line from the bottom, for "towards" read "from."

Ibid., three lines from the bottom, after "Italy," add "to take the place of the ascending current from the Sahara."

In the Hurricane Chart opposite to page 61, by Sir John Reid, (erroneously printed "Reade,") the dotted line traversing the tinted portion denotes the course of the ship "Castries," which, on the 23rd of August, 1837, and therefore almost cotemporaneously with the hurricane delineated in the untinted part of the map, was involved in a smaller vorticose storm, on the spot designated by the concentric lines.

Publications by the Author.

A DESCRIPTION OF ACTIVE AND EXTINCT VOLCANOS, OF EARTHQUAKES, AND OF THERMAL SPRINGS; With remarks on their Causes, Products, and Influence on the Condition of the Globe. Second Edition, greatly enlarged. With Twelve Maps and Plates.
London: TAYLOR and FRANCIS. 1848.

BRIEF REMARKS ON THE CORRELATION OF THE NATURAL SCIENCES. Drawn up with reference to the scheme for the extension and the better management of the Studies of the University.
Oxford: Printed and Published by J. VINCENT. 1848.

A POPULAR GUIDE TO THE BOTANIC GARDEN OF OXFORD, AND TO THE FIELDING HERBARIUM ANNEXED TO IT. Second Edition. *Sold only at the Garden.*

AN INTRODUCTION TO THE ATOMIC THEORY. Second Edition, greatly enlarged. 1852. Published at the OXFORD UNIVERSITY PRESS.

CAN PHYSICAL SCIENCE OBTAIN A HOME IN AN ENGLISH UNIVERSITY? An inquiry suggested by some remarks contained in a late number of the "Quarterly Review."
Oxford: VINCENT. 1853.

ON THE IMPORTANCE OF THE STUDY OF CHEMISTRY AS A BRANCH OF EDUCATION FOR ALL CLASSES. A Lecture delivered at the Royal Institution of Great Britain, 1854.
London: J. W. PARKER, West Strand.

LECTURES ON ROMAN HUSBANDRY, delivered before the University of Oxford; comprehending such an account of the System of Agriculture, the Treatment of Domestic Animals, the Horticulture, &c., pursued in Ancient Times, as may be collected from the *Scriptores rei Rusticæ*, the Georgics of Virgil, and other Classical Authorities, with Notices of the Plants mentioned in Columella and Virgil. 8vo., cloth, reduced to 6s.
Oxford and London: J. H. and J. PARKER.

CORRECTIONS AND ADDITIONS.

Page 28. The foot-note ought to stand thus:—"Hooker states, that the most southern latitude in which Tree-ferns have been met with is in the north of New Zealand, south latitude 44°. Alsophila australis and Dicksonia antarctica, both of them Tree-ferns, are likewise found in North Tasmania at about the same latitude, the latter indeed most abundantly; and even in lat. 50° 5', in the Lord Auckland group, the Aspidium venustum becomes subarborescent near the level of the sea. See Flora Tasmaniæ."

P. 58, sixteenth line from the bottom, for "towards" read "from."

Ibid., three lines from the bottom, after "Italy," add "to take the place of the ascending current from the Sahara."

In the Hurricane Chart opposite to page 61, by Sir John Reid, (erroneously printed "Reade,") the dotted line traversing the tinted portion denotes the course of the ship "Castries," which, on the 23rd of August, 1837, and therefore almost simultaneously with the hurricane delineated in the untinted part of the map, was involved in a similar vorticose storm, on the spot designated by the concentric lines.

P. 74. After the words in the first line, "for the warm ones," add as follows: "Accordingly the former class would predominate in colder regions, were it not, that in proportion as we recede from the equator there is an increase in the number of Grasses and of Sedges, which, as they die down in winter, are capable of withstanding its rigour, and thus by their abundance serve to give a preponderance to the monocotyledonous flora of high latitudes, notwithstanding the want of arborescent species possessing that structure."

In p. 79, line 11, for "arises" read "possibly may arise."

In p. 80, line 29, for "the Mesembryanthemums and Sempervivums," read "the succulent plants, which are there met with; species either allied to the Mesembryanthemums, (Reaumuriæ, &c.), and hence confounded with them by some travellers, although only two genuine species of this extensive tribe are mentioned as occurring in the East, and none in Arabia especially; or else belonging to the family Crassulaceæ, such as the Cotyledons or Kalanchoes, noticed by Forskal amongst the productions of that arid region."

In p. 82 cancel the foot-note.

Page 84. What is here stated as to the competency of the seeds, &c., of plants to support life for an indefinite period, applies, of course, only to those which contain, in addition to starch and sugar, a certain per-centage of some nitrogenous principle. Thus we are assured by Dr. Forbes Watson, that in India, where, in the absence of rice, various cereals (some of them enumerated in p. 94) more deficient in nitrogen than wheat are cultivated

CORRECTIONS AND ADDITIONS.

for food, a certain proportion of some kind of pulse is a constant addition to the dietary of the native population, and that the quantity which they habitually employ is exactly that, which science has found to be necessary, in order that the mixture should contain the same proportion of carbon (C) to nitrogen (N) as exists in Wheat. Thus

Proportion of C. to N.	Cereal employed as food.	Species of Pulse added to make up the quota.			
		Cicer Arietinum C. to N. as 3·8 to 1.	Calamus indious C. to N. as 3·2 to 1.	Dolichos uniflorus, Dolichos sinensis, Phaseolus aconitifolius C. to N. as 2·7 to 1.	Phaseolus Mungo C. to N. 2·5 to 1.
13·7 to 1	Cynosurus coracana 100 parts	114	75	59	62
11·1 to 1	Rice	92	63	47	50
8·5 to 1	Sorghum vulgare	57	39	29	31
8·1 to 1	Maize . . .	57	39	29	31
7·6 to 1	Holcus spicatus	41	28	21	22
6·4 to 1	Panicum miliaceum . . .	12	8	6	6

In p. 88, line 22, for "so malignant as to be employed by the Indians for poisoning their arrows," read "of a most malignant nature."

Ibid., line 26, for "Cassarine" read "Cassaripe," and add, "I may state, however, on the authority of the Archbishop of Dublin, that there is a variety of the Cassava, not a distinct species, which possesses no poisonous properties at all, and is roasted or boiled like a potato, without any previous preparation."

P. 89, line 6, for "the principal article of food," read "cultivated as an article of food."

Ibid., line 12, for "so that it is not met with at any great distance from the ocean," read "but it is met with nevertheless both in the interior of the Indian Peninsula, and high up the Niger."

Ibid., line 23, for "extensively" read "stated by Meyen to be."

Ibid., line 29, for "the only material which" read "the material which chiefly."

Page 91, line 12, for "New Zealand" read "the Sandwich Islands."

Ibid., line 14, after the words "in that island," add "and likewise in New Zealand, as well as in other parts of Polynesia, where it has been introduced, though not indigenous to the soil."

Page 92, line 10, for "sea-weed" read "lichen."

Ibid., line 13, instead of the sentence, "It must, however, go through some process of maceration or cooking before it can be fit for digestion," read as follows:—"The specimen of it in my possession is so hard and tough as to be utterly unfit for digestion, unless it went through some process of maceration, but I am assured by Dr. Joseph Hooker, that when first dug up it is soft, and therefore can be eaten raw."

In p. 94, six lines from the bottom, after "Amaranthus fariniferus," add, "according to Meyen, although Dr. Joseph Hooker questions the fact;

reminding me also, that I had omitted to enumerate Buckwheat, 'Polygonum fagopyrum,' as a plant extensively cultivated in those regions for food."

Page 122, after line 21, add as follows:—

The Archbishop of Dublin informs me "that the red-flowering Ribes (R. sanguineum) although when first brought over it flowered freely, did not then produce fruit, but that after remaining some years in the country it was observed to bear here and there a berry, and has continued progressing in this respect, until it ended by becoming loaded with fruit every year."

"Two other species of Ribes also, the aureum, and the prickly species, (speciosum?), he says, have begun after a few years' sojourn in this country to bear a few berries."

Such facts as these might lead us to infer, that within certain limits a change can be effected in the constitution of a plant, by which it becomes accommodated to new conditions of climate, but it does not therefore follow, that this can go on indefinitely, so as to afford encouragement to the hopes of those who espouse the theory of acclimatisation.

The Archbishop also cites some instances, in which the mere difference, small as it may seem, between the climate of countries so contiguous as England and Ireland, exercises a marked influence upon the flowering of a plant.

"Thus," he says, "the Buddlea, 'Buddleja globosa,' although it flowers freely in England, so seldom produces fruit, that in Suffolk an instance of the kind which occurred in his garden was regarded as a great curiosity, whereas in Ireland the plant is loaded with seed-vessels every year.

"In Suffolk the Archbishop had in his garden a Laburnum-tree, (Cytisus Laburnum,) one of whose branches, about as thick as the finger, swelled out towards the extremity nearly to the thickness of one's wrist, and from the bulging part of which a dozen or more luxuriant shoots pushed out.

"This was regarded by the London Horticultural Society as a great curiosity, and yet in Ireland nearly half the Laburnums send forth such branches."

Appendix II. p. 142. Add to the list of plants cultivated in the open air by Mr. R. (R. W.) Fox at Penjerrick, (printed by mistake Pengellert,) near Falmouth, Mitraria coccinea, and Rhododendron Edgworthii, both of which have flowered freely. At Grove Hill, near Falmouth, the orange-flowering Datura (Brugmansia lutea?) and the Clianthus, bear abundance of flowers every year.

Publications by the Author.

A DESCRIPTION OF ACTIVE AND EXTINCT VOLCANOS, OF EARTHQUAKES, AND OF THERMAL SPRINGS; With remarks on their Causes, Products, and Influence on the Condition of the Globe. Second Edition, greatly enlarged. With Twelve Maps and Plates.
   London: TAYLOR and FRANCIS. 1848.

BRIEF REMARKS ON THE CORRELATION OF THE NATURAL SCIENCES. Drawn up with reference to the scheme for the extension and the better management of the Studies of the University.
   Oxford: Printed and Published by J. VINCENT. 1848.

A POPULAR GUIDE TO THE BOTANIC GARDEN OF OXFORD, AND TO THE FIELDING HERBARIUM ANNEXED TO IT. Second Edition. *Sold only at the Garden.*

AN INTRODUCTION TO THE ATOMIC THEORY. Second Edition, greatly enlarged. 1852. Published at the OXFORD UNIVERSITY PRESS.

CAN PHYSICAL SCIENCE OBTAIN A HOME IN AN ENGLISH UNIVERSITY? An inquiry suggested by some remarks contained in a late number of the "Quarterly Review."
   Oxford: VINCENT. 1853.

ON THE IMPORTANCE OF THE STUDY OF CHEMISTRY AS A BRANCH OF EDUCATION FOR ALL CLASSES. A Lecture delivered at the Royal Institution of Great Britain, 1854.
   London: J. W. PARKER, West Strand.

LECTURES ON ROMAN HUSBANDRY, delivered before the University of Oxford; comprehending such an account of the System of Agriculture, the Treatment of Domestic Animals, the Horticulture, &c., pursued in Ancient Times, as may be collected from the *Scriptores rei Rusticæ*, the Georgics of Virgil, and other Classical Authorities, with Notices of the Plants mentioned in Columella and Virgil. 8vo., cloth, reduced to 6s.
   Oxford and London: J. H. and J. PARKER.

A List of Books
RECENTLY PUBLISHED BY
JOHN HENRY AND JAMES PARKER,
OXFORD, AND 377, STRAND, LONDON.

NEW THEOLOGICAL WORKS.

REV. JOHN KEBLE.

THE LIFE OF THE RIGHT REVEREND FATHER IN GOD, THOMAS WILSON, D.D., Lord Bishop of Sodor and Man. Compiled, chiefly from Original Documents, by the Rev. JOHN KEBLE, M.A., Vicar of Hursley. In Two Parts, 8vo, price 21s.

Forming Vols. 87 and 88 of the Anglo-Catholic Library.

BISHOP WILSON.

THE WORKS OF THE RIGHT REVEREND FATHER IN GOD, THOMAS WILSON, D.D., Lord Bishop of Sodor and Man.—Vol. VII., containing PAROCHIALIA, with other TRACTS and FRAGMENTS, and a GENERAL INDEX. 8vo., cloth, price 7s.

Forming Vol. 86 of the Anglo-Catholic Library.

THE LORD BISHOP OF OXFORD.

SERMONS PREACHED BEFORE THE UNIVERSITY OF OXFORD: SECOND SERIES, from MDCCCXLVII. to MDCCCLXII. 8vo., cloth, 10s. 6d.

THE ORDINATION SERVICE. ADDRESSES ON THE QUESTIONS TO THE CANDIDATES FOR ORDINATION. By the Right Rev. the LORD BISHOP OF OXFORD. *Fourth Edition.* Crown 8vo., cloth, 6s.

FELLOWSHIP IN JOY AND SORROW. A Sermon preached in Her Majesty's Royal Chapel in Windsor Castle, on the Sunday preceding the Marriage of H. R. H. the PRINCE OF WALES, March 8, 1863. By SAMUEL, LORD BISHOP OF OXFORD, Lord High Almoner to the Queen; Chancellor of the Most Noble Order of the Garter. 8vo., in wrapper, price 6d. (Published by Her Majesty's command.)

REV. P. FREEMAN.

THE PRINCIPLES OF DIVINE SERVICE; or, An Inquiry concerning the True Manner of Understanding and Using the Order for Morning and Evening Prayer, and for the Administration of the Holy Communion in the English Church. By the Rev. PHILIP FREEMAN, M.A., Vicar of Thorverton, Prebendary of Exeter, and Examining Chaplain to the Lord Bishop of Exeter. 2 vols., 8vo., cloth, price 1l. 4s. *The Second Edition of Vol. I. is now ready.*

".... It will take its rank with Mabillon's Gallican Liturgies, or Bona's great work, while in style it soars high above them. We congratulate the Church of England on receiving such a work from one of her priests."—*Christian Remembrancer.*

OXFORD LENTEN SERMONS.

THE OXFORD LENTEN SERMONS FOR 1863, preached in the Church of St. Mary-the-Virgin. By the LORD ARCHBISHOP OF YORK; Rev. PROF. MANSEL; Rev. DR. WORDSWORTH; Rev. T. L. CLAUGHTON, M.A.; Rev. DR. STANLEY; Rev. T. T. CARTER, M.A.; the LORD BISHOP OF LONDON; Rev. J. R. WOODFORD, M.A.; Rev. DR. PUSEY; Rev. D. MOORE, M.A.; Rev. DR. MAGEE; Very Rev. DR. ALFORD. 8vo., cloth, 7s. 6d.

THEOLOGICAL WORKS, (continued).

THE LATE DR. WILLIAMS.
SERMONS preached before the University of Oxford, and in Winchester Cathedral, by the late DAVID WILLIAMS, D.C.L., Warden of New College, Oxford, and Canon of Winchester; formerly Head Master of Winchester College. WITH A BRIEF MEMOIR OF THE AUTHOR. 8vo., cloth, 10s. 6d.

REV. R. PAYNE SMITH.
THE AUTHENTICITY AND MESSIANIC INTERPRETATION OF THE PROPHECIES OF ISAIAH vindicated in a Course of Sermons preached before the University of Oxford, by the Rev. R. PAYNE SMITH, M.A., Sub-Librarian of the Bodleian Library, and Select Preacher. 8vo., cloth, 10s. 6d.

REV. C. A. HEURTLEY, D.D.
THE FORM OF SOUND WORDS: HELPS TOWARDS HOLDING IT FAST: Seven Sermons preached before the University of Oxford, on some Important Points of Faith and Practice.
HINDRANCES TO SUCCESS IN PREACHING.
THE FORM OF SOUND WORDS.
THE INSPIRATION OF SCRIPTURE.
THE CONNECTION BETWEEN BAPTISM AND
SANCTIFICATION.
CONFESSION AND ABSOLUTION.
THE DOCTRINE OF THE ATONEMENT.
THE LORD'S DAY.
By CHARLES A. HEURTLEY, D.D., Margaret Professor of Divinity, and Canon of Christ Church. 8vo., cloth, 7s. 6d.

ARCHDEACON CHURTON.
A MEMOIR OF THE LATE JOSHUA WATSON, ESQ. By the Venerable Archdeacon CHURTON. *A new and cheaper Edition, with Portrait.* Crown 8vo., cloth, 7s. 6d.

REV. DR. MOBERLY.
SERMONS ON THE BEATITUDES, with others mostly preached before the University of Oxford; to which is added a Preface relating to the recent volume of "Essays and Reviews." By the Rev. GEORGE MOBERLY, D.C.L., Head Master of Winchester College. *Second Edition.* 8vo., price 10s. 6d.
The Preface separately, price 2s.

REV. E. B. PUSEY, D.D.
THE MINOR PROPHETS; with a Commentary Explanatory and Practical, and Introductions to the Several Books. By the Rev. E. B. PUSEY, D.D., Regius Professor of Hebrew, and Canon of Christ Church. Part III. Amos vi. 6 to end—Obadiah—Jonah—Micah i. 12. 4to., sewed, 5s.
Lately published, 4to., sewed, 5s.
Part I. Hosea—Joel, Introduction. Part II. Joel, Introduction—Amos vi. 6.
THE COUNCILS OF THE CHURCH, from the Council of Jerusalem, A.D. 51, to the Council of Constantinople, A.D. 381; chiefly as to their Constitution, but also as to their Objects and History. By the Rev. E. B. PUSEY, D.D., Regius Professor of Hebrew; Canon of Christ Church; late Fellow of Oriel College. 8vo., 10s. 6d.

REV. J. W. BURGON.
INSPIRATION AND INTERPRETATION. Seven Sermons preached before the University of Oxford; with an Introduction, being an answer to a Volume entitled "Essays and Reviews." By the Rev. JOHN W. BURGON, M.A., Fellow of Oriel College, and Select Preacher. 8vo., cloth, 14s.

REV. WILLIAM BRIGHT.
A HISTORY OF THE CHURCH, from the EDICT of MILAN, A.D. 313, to the COUNCIL of CHALCEDON, A.D. 451. By WILLIAM BRIGHT, M.A., Fellow of University College, Oxford; late Professor of Ecclesiastical History in the Scottish Church. Post 8vo., price 10s. 6d.
ANCIENT COLLECTS AND OTHER PRAYERS, Selected for Devotional Use from various Rituals, with an Appendix on the Collects in the Prayer-book. By WILLIAM BRIGHT, M.A., Fellow of University College, Oxford, Author of "A History of the Church," &c. *Second Edition,* enlarged, printed in red and black, on toned paper, antique cloth, 5s.

DR. ELVEY.
THE PSALTER, or Canticles and Psalms of David, Pointed for Chanting, upon a New Principle; with Explanations and Directions. By the late STEPHEN ELVEY, Mus. Doc., Organist of New and St. John's Colleges, and Organist and Choragus to the University of Oxford. *Second Edition,* 8vo., cloth, price 7s. 6d.

ST. JOHN CHRYSOSTOM.
S. JOANNIS CHRYSOSTOMI INTERPRETATIO OMNIUM EPISTOLARUM PAULINARUM PER HOMILIAS FACTA. Tomus VI. IN EPISTOLAS AD TIMOTHEUM, TITUM, ET PHILEMONEM. 8vo., cloth, price 10s. 6d.

—————————————————— TOMUS VII. Continens Homilias in Epistolam ad Hebraeos, et Indices. 8vo., cloth, 12s.

ST. JUSTIN MARTYR.
THE WORKS OF S. JUSTIN MARTYR, Translated, with Notes and Indices. 8vo., cloth, price 8s.

ST. ANSELM.
CUR DEUS HOMO, or WHY GOD WAS MADE MAN; by ST. ANSELM. Translated. *Second Edition.* Fcap. 8vo., 2s. 6d.

MEDITATIONS AND SELECT PRAYERS, by ST. ANSELM, formerly Archbishop of Canterbury. Edited by E. B. PUSEY, D.D. Fcap. 8vo., 5s.

REPLIES TO ESSAYS AND REVIEWS.
REPLIES TO **"ESSAYS AND REVIEWS."** By

I. THE REV. E. M. GOULBURN, D.D.
II. THE REV. H. J. ROSE, B.D.
III. THE REV. C. A. HEURTLEY, D.D.
IV. THE REV. W. J. IRONS, D.D.
V. THE REV. G. ROBISON, LL.D.
VI. THE REV. A. W. HADDAN, B.D.
VII. THE REV. CHR. WORDSWORTH, D.D.

With a Preface by the Lord Bishop of Oxford; and Letters from the Radcliffe Observer and the Reader in Geology in the University of Oxford.—*Second Edition, with a Note by Professor Owen.* 8vo., cloth, 12s.

REV. JOHN WALKER.
THE SUFFERINGS OF THE CLERGY DURING THE GREAT REBELLION. By the Rev. JOHN WALKER, M.A., sometime of Exeter College, Oxford, and Rector of St. Mary Major, Exeter. Epitomised by the Author of "The Annals of England." Fcap. 8vo., cloth, 5s.

REV. W. H. KARSLAKE.
AN EXPOSITION OF THE LORD'S PRAYER, Devotional, Doctrinal, and Practical; with Four Preliminary Dissertations, and an Appendix of Extracts from Writers on the Prayer for Daily Use. By the Rev. W. H. KARSLAKE, Fellow and sometime Tutor of Merton College, Oxford. 8vo., cloth, 7s. 6d.

REV. T. LATHBURY.
A HISTORY OF THE BOOK OF COMMON PRAYER, AND OTHER AUTHORIZED BOOKS, from the Reformation; and an Attempt to ascertain how the Rubrics, Canons, and Customs of the Church have been understood and observed from the same time: with an Account of the State of Religion in England from 1640 to 1660. By the Rev. THOMAS LATHBURY, M.A., Author of "A History of the Convocation," "The Nonjurors," &c. *Second Edition, with an Index.* 8vo., 10s. 6d.

THEOLOGICAL WORKS, (continued).

THE LATE REV. H. NEWLAND.

A NEW CATENA ON ST. PAUL'S EPISTLES.—A PRACTICAL AND EXEGETICAL COMMENTARY ON THE EPISTLES OF ST. PAUL TO THE EPHESIANS, AND THE PHILIPPIANS; in which are exhibited the Results of the most learned Theological Criticisms, from the Age of the Early Fathers down to the Present Time. Edited by the late Rev. HENRY NEWLAND, M.A., Vicar of St. Mary Church, Devon, and Chaplain to the Bishop of Exeter. 8vo., cl., 12s.

REV. H. DOWNING.

SHORT NOTES ON ST. JOHN'S GOSPEL, intended for the Use of Teachers in Parish Schools, and other Readers of the English Version. By HENRY DOWNING, M.A., Incumbent of St. Mary's, Kingswinford. Fcap. 8vo., cloth, 2s. 6d.

SHORT NOTES ON THE ACTS OF THE APOSTLES, intended for the use of Teachers in Parish Schools, and other Readers of the English Version. By the same Author. Fcap. 8vo., cloth, 2s.

REV. L. P. MERCIER.

CONSIDERATIONS RESPECTING A FUTURE STATE. By the Rev. LEWIS P. MERCIER, M.A., University College, Oxford. Fcap. 8vo., 4s.

REV. J. M. NEALE.

A HISTORY OF THE SO-CALLED JANSENIST CHURCH OF HOLLAND; with a Sketch of its Earlier Annals, and some Account of the Brothers of the Common Life. By the Rev. J. M. NEALE, M.A., Warden of Sackville College. 8vo., cloth, reduced to 5s.

REV. R. W. MORGAN.

ST. PAUL IN BRITAIN; or, THE ORIGIN OF BRITISH AS OPPOSED TO PAPAL CHRISTIANITY. By the Rev. R. W. MORGAN, Perpetual Curate of Tregynon, Montgomeryshire, Author of "Verities of the Church," "The Churches of England and Rome," "Christianity and Infidelity intellectually contrasted," &c. Crown 8vo., cloth, 4s.

THE LATE DEAN OF FERNS.

THE LIFE AND CONTEMPORANEOUS CHURCH HISTORY OF ANTONIO DE DOMINIS, Archbishop of Spalatro, which included the Kingdoms of Dalmatia and Croatia; afterwards Dean of Windsor, Master of the Savoy, and Rector of West Ilsley in the Church of England, in the reign of James I. By the late HENRY NEWLAND, D.D., Dean of Ferns. 8vo., cloth lettered, reduced to 7s.

REV. J. DAVISON.

DISCOURSES ON PROPHECY, in which are considered its Structure, Use, and Inspiration; being the substance of Twelve Sermons preached in the Chapel of Lincoln's-Inn, by JOHN DAVISON, B.D. Sixth and cheaper Edition. 8vo., cloth, 9s.

REV. DR. SEWELL.

CHRISTIAN VESTIGES OF CREATION. By WILLIAM SEWELL, D.D., late Professor of Moral Philosophy in the University of Oxford. Post 8vo., cloth, price 4s. 6d.

SERMONS, &c.

ILLUSTRATIONS OF FAITH. EIGHT PLAIN SERMONS, by a Writer in the "Tracts for the Christian Seasons:"—Abel; Enoch; Noah; Abraham; Isaac, Jacob, and Joseph; Moses; The Walls of Jericho; Conclusions. Fcap. 8vo., cloth, 2s. 6d.

Uniform, and by the same Author,

PLAIN SERMONS ON THE BOOK OF COMMON PRAYER. Fcap. 8vo., cloth, 5s.

HISTORICAL AND PRACTICAL SERMONS ON THE SUFFERINGS AND RESURRECTION OF OUR LORD. 2 vols., Fcap. 8vo., cloth, 10s.

SERMONS ON NEW TESTAMENT CHARACTERS. Fcap. 8vo., 4s.

CHRISTIAN SEASONS.—Short and Plain Sermons for every Sunday and Holyday throughout the Year. Edited by the late Bishop of Grahamstown. 4 vols., Fcap. 8vo., cloth, 16s.

——— A Second Series of Sermons for the Christian Seasons. Uniform with the above. 4 vols., Fcap. 8vo., cloth, 16s.

ARMSTRONG'S PAROCHIAL SERMONS. Parochial Sermons, by JOHN ARMSTRONG, D.D., late Lord Bishop of Grahamstown. A New Edition, Fcap. 8vo., cloth, 5s.

ARMSTRONG'S SERMONS FOR FASTS AND FESTIVALS. A new Edition. Fcap. 8vo., 5s.

PAROCHIAL SERMONS, by the Rev. HENRY W. BURROWS, B.D., Perpetual Curate of Christ Church, St. Pancras. Fcap. 8vo., cloth, 6s.

——————— Second Series. Fcap. 8vo., cloth, 5s.

SERMONS PREACHED FOR THE MOST PART IN THE CHURCHES OF ST. MARY AND ST. MATTHIAS, RICHMOND, SURREY. By CHARLES WELLINGTON FURSE, M.A., of Balliol College; Curate of Christ Church, St. Pancras; and formerly Lecturer of St. George's Chapel, Windsor. Post 8vo., cloth, 6s.

SHORT SERMONS FOR FAMILY READING. Ninety Short Sermons for Family Reading, following the course of the Christian Seasons. By the Author of "A Plain Commentary on the Gospels." 2 vols., cloth, 8s.

SERMONS PREACHED BEFORE THE UNIVERSITY OF OXFORD, and in other places. By the late Rev. C. MARRIOTT, Fellow of Oriel College, Oxford. 12mo., cloth, 6s.

——— Volume the Second. 12mo., cloth, 7s. 6d.

LECTURES ON THE EPISTLE OF ST. PAUL TO THE ROMANS. By the late Rev. C. MARRIOTT, B.D., Fellow of Oriel College, Oxford. Edited by his Brother, the Rev. JOHN MARRIOTT. 12mo., cloth, 6s.

REFLECTIONS IN A LENT READING OF THE EPISTLE TO THE ROMANS. By the late Rev. C. MARRIOTT. Fcap. 8vo., cloth, 3s.

Works of the Standard English Divines,

PUBLISHED IN THE LIBRARY OF ANGLO-CATHOLIC THEOLOGY,

AT THE FOLLOWING PRICES IN CLOTH.

ANDREWES' (BP.) COMPLETE WORKS. 11 vols., 8vo., £3 7s.
    THE SERMONS. (Separate.) 5 vols., £1 15s.
BEVERIDGE'S (BP.) COMPLETE WORKS. 12 vols., 8vo., £4 4s.
    THE ENGLISH THEOLOGICAL WORKS. 10 vols., £3 10s.
BRAMHALL'S (ABP.) WORKS, WITH LIFE AND LETTERS, &c. 5 vols., 8vo., £1 15s.
BULL'S (BP.) HARMONY ON JUSTIFICATION. 2 vols., 8vo., 10s.
————— DEFENCE OF THE NICENE CREED. 2 vols., 10s.
————— JUDGMENT OF THE CATHOLIC CHURCH. 5s.
COSIN'S (BP.) WORKS COMPLETE. 5 vols., 8vo., £1 10s.
CRAKANTHORP'S DEFENSIO ECCLESIÆ ANGLICANÆ. 8vo., 7s.
FRANK'S SERMONS. 2 vols., 8vo., 10s.
FORBES' CONSIDERATIONES MODESTÆ. 2 vols., 8vo., 12s.
GUNNING'S PASCHAL, OR LENT FAST. 8vo., 6s.
HAMMOND'S PRACTICAL CATECHISM. 8vo., 5s.
————— MISCELLANEOUS THEOLOGICAL WORKS. 5s.
————— THIRTY-ONE SERMONS. 2 Parts. 10s.
HICKES'S TWO TREATISES ON THE CHRISTIAN PRIESTHOOD. 3 vols., 8vo., 15s.
JOHNSON'S (JOHN) THEOLOGICAL WORKS. 2 vols., 8vo., 10s.
————— ENGLISH CANONS. 2 vols., 12s.
LAUD'S (ABP.) COMPLETE WORKS. 6 vols., (8 Parts,) 8vo., £2 10s.
L'ESTRANGE'S ALLIANCE OF DIVINE OFFICES. 8vo., 6s.
MARSHALL'S PENITENTIAL DISCIPLINE. 8vo., 4s.
NICHOLSON'S (BP.) EXPOSITION OF THE CATECHISM. (This volume cannot be sold separate from the complete set.)
OVERALL'S (BP.) CONVOCATION-BOOK OF 1606. 8vo., 5s.
PEARSON'S (BP.) VINDICIÆ EPISTOLARUM S. IGNATII. 2 vols. 8vo., 10s.
THORNDIKE'S (HERBERT) THEOLOGICAL WORKS COMPLETE. 6 vols., (10 Parts,) 8vo., £2 10s.
WILSON'S (BP.) WORKS COMPLETE. With LIFE, by Rev. J. KEBLE. 7 vols., (8 Parts,) 8vo., £3 3s.

A complete set, £25.

NEW DEVOTIONAL WORKS.

DAILY STEPS TOWARDS HEAVEN; or, Practical Thoughts on the Gospel History, and especially on the Life and Teaching of our Lord Jesus Christ, for every day in the year, according to the Christian Seasons. With Titles and Characters of Christ; and a Harmony of the Four Gospels. *Eleventh Edition.* 32mo., roan, 2s. 6d.; morocco, 4s. 6d.

———————— LARGE-TYPE EDITION, square crown 8vo., cloth, price 5s.

THE PASTOR IN HIS CLOSET; or, A Help to the Devotions of the Clergy. By JOHN ARMSTRONG, D.D., late Lord Bishop of Grahamstown. *Third Edition.* Fcap. 8vo., cloth, 2s.

DAILY SERVICES FOR CHRISTIAN HOUSEHOLDS, compiled and arranged by the Rev. H. STOBART, M.A. 18mo., paper, 1s.; cloth, 1s. 4d.

THOUGHTS DURING SICKNESS. By the Author of "The Doctrine of the Cross," and "Devotions for the Sick Room." *Second Edition.* Price 2s. 6d.

BREVIATES FROM HOLY SCRIPTURE, arranged for use by the Bed of Sickness. By the Rev. G. ARDEN, M.A., Rector of Winterborne-Came; Domestic Chaplain to the Right Hon. the Earl of Devon; Author of "A Manual of Catechetical Instruction." Fcap. 8vo. *Second Edition.* 2s.

THE CURE OF SOULS. By the Rev. G. ARDEN, M.A. Fcap. 8vo., 2s. 6d.

PRECES PRIVATÆ in studiosorum gratiam collectæ et regia auctoritate approbatæ: anno MDLXVIII. *Londini* editæ: ad vetera exemplaria denuo recognitæ. Ed. C. MARRIOTT. 16mo., cloth, 6s.

OXFORD SERIES OF DEVOTIONAL WORKS.

THE IMITATION OF CHRIST.
FOUR BOOKS. By Thomas À KEMPIS. A new Edition, revised, handsomely printed on tinted paper in Fcap. 8vo., with Vignettes and red borders, cl., 5s.; antique calf, red edges, 10s. 6d.

LAUD'S DEVOTIONS.
THE PRIVATE DEVOTIONS of Dr. WILLIAM LAUD, Archbishop of Canterbury, and Martyr. A new and revised Edition, with Translations to the Latin Prayers, handsomely printed with Vignettes and red lines. Fcap. 8vo., antique cloth, 5s.

WILSON'S SACRA PRIVATA.
THE PRIVATE MEDITATIONS, DEVOTIONS, and PRAYERS of the Right Rev. T. WILSON, D.D., Lord Bishop of Sodor and Man. Now first printed entire. From the Original Manuscripts. Fcap. 8vo., 6s.

ANDREWES' DEVOTIONS.
DEVOTIONS. By the Right Rev. Father in God, LAUNCELOT ANDREWES, Translated from the Greek and Latin, and arranged anew. Fcap. 8vo., 5s.; morocco, 8s.; antique calf, red edges, 10s. 6d.

SPINCKES' DEVOTIONS.
TRUE CHURCH OF ENGLAND MAN'S COMPANION IN THE CLOSET; or, a complete Manual of Private Devotions, collected from the Writings of eminent Divines of the Church of England. Sixteenth Edition, corrected. Fcap. 8vo., floriated borders, cloth, antique, 4s.

The above set of 5 Volumes, in neat grained calf binding, £2 2s.

TAYLOR'S HOLY LIVING.
THE RULE AND EXERCISES OF HOLY LIVING. By BISHOP JEREMY TAYLOR. In which are described the means and instruments of obtaining every virtue, and the remedies against every vice. *In antique cloth binding,* 4s.

TAYLOR'S HOLY DYING.
THE RULE AND EXERCISES OF HOLY DYING. By BISHOP JEREMY TAYLOR. In which are described the means and instruments of preparing ourselves and others respectively for a blessed death, &c. *In antique cloth binding,* 4s.

ANCIENT COLLECTS.
Lately published. Vide p. 2.

CHURCH POETRY.

THE AUTHOR OF "THE CHRISTIAN YEAR."

THE CHRISTIAN YEAR. Thoughts in verse for the Sundays and Holydays throughout the Year. *Imperial Octavo*, with Illuminated Titles,—Cloth, 1l. 5s.; morocco, 1l. 11s. 6d.; best morocco, 2l. 2s. *Octavo Edition*,—Large type, cloth, 10s. 6d.; morocco by Hayday, 21s.; antique calf, 18s. *Foolscap Octavo Edition*,—Cloth, 7s. 6d.; morocco, 10s. 6d.; morocco by Hayday, 15s.; antique calf, 12s. *32mo. Edition*,—Cloth, 3s. 6d.; morocco, plain, 5s.; morocco by Hayday, 7s. *Cheap Edition*,—Cloth, 1s. 6d.; bound, 2s.

LYRA INNOCENTIUM. Thoughts in Verse for Christian Children. *Foolscap Octavo Edition*,—Cloth, 7s. 6d.; morocco, plain, 10s. 6d.; morocco by Hayday, 15s.; antique calf, 12s. *18mo. Edition*,—Cloth, 6s.; morocco, 8s. 6d. *32mo. Edition*,—Cloth, 3s. 6d.; morocco, plain, 5s.; morocco by Hayday, 7s. *Cheap Edition*,—Cloth, 1s. 6d.; bound, 2s.

THE AUTHOR OF "THE CATHEDRAL."

THE CATHEDRAL. Foolscap 8vo., cloth, 7s. 6d.; 32mo., with Engravings, 4s. 6d.

THOUGHTS IN PAST YEARS. *The Sixth Edition*, with several new Poems, 32mo., cloth, 4s. 6d.

THE BAPTISTERY; or, The Way of Eternal Life. 32mo., cloth, 3s. 6d.

The above Three Volumes uniform, 32mo., neatly bound in morocco, 18s.

THE CHRISTIAN SCHOLAR. Foolscap 8vo., 10s. 6d.; 32mo., cloth, 4s. 6d.

THE SEVEN DAYS; or, The Old and New Creation. *Second Edition*, Foolscap 8vo., 7s. 6d.

MORNING THOUGHTS. By a CLERGYMAN. Suggested by the Second Lessons for the Daily Morning Service throughout the year. 2 vols. Foolscap 8vo., cloth, 5s. each.

THE CHILD'S CHRISTIAN YEAR. Hymns for every Sunday and Holyday throughout the year. *Cheap Edition*, 18mo., cloth, 1s.

COXE'S CHRISTIAN BALLADS. Foolscap 8vo., cloth, 3s. Also selected Poems in a packet, sewed, 1s.

FLORUM SACRA. By the Rev. G. HUNT SMYTTAN. *Second Edition*, 16mo., 1s.

PROFESSOR GOLDWIN SMITH.

IRISH HISTORY AND IRISH CHARACTER. By GOLDWIN SMITH. *Second Edition.* Post 8vo., price 5s.

Uniform with the above.

THE EMPIRE. A SERIES OF LETTERS PUBLISHED IN "THE DAILY NEWS," 1862, 1863. By GOLDWIN SMITH. Post 8vo., cloth, price 6s.

PROFESSOR WILLIS.

FACSIMILE OF THE SKETCH-BOOK OF WILARS DE HONECORT, an Architect of the Thirteenth Century. With Commentaries and Descriptions by MM. LASSUS and QUICHERAT. Translated and Edited, with many additional Articles and Notes, by the Rev. ROBERT WILLIS, M.A., F.R.S., Jacksonian Professor at Cambridge, &c. With 64 Facsimiles, 10 Illustration Plates, and 43 Woodcuts. Royal 4to., cloth, 2l. 10s.

The English letterpress separate, for the purchasers of the French edition, 4to., 15s.

JOHN HENRY PARKER.

AN INTRODUCTION TO THE STUDY OF GOTHIC ARCHITECTURE. By JOHN HENRY PARKER, F.S.A. *Second Edition*, Revised and Enlarged, with 170 Illustrations, and a Glossarial Index. Fcap. 8vo., cloth lettered, price 5s.

AN ATTEMPT TO DISCRIMINATE THE STYLES OF ARCHITECTURE IN ENGLAND, FROM THE CONQUEST TO THE REFORMATION: WITH A SKETCH OF THE GRECIAN AND ROMAN ORDERS. By the late THOMAS RICKMAN, F.S.A. Sixth Edition, with considerable Additions, chiefly Historical, by JOHN HENRY PARKER, F.S.A., and numerous Illustrations by O. Jewitt. 8vo., cloth, price 1l. 1s.

JOHN HEWITT.

ANCIENT ARMOUR AND WEAPONS IN EUROPE. By JOHN HEWITT, Member of the Archæological Institute of Great Britain. Vols. II. and III., comprising the Period from the Fourteenth to the Seventeenth Century, completing the work, 1l. 12s. Also Vol. I., from the Iron Period of the Northern Nations to the end of the Thirteenth Century, 18s. The work complete, 3 vols., 8vo., 2l. 10s.

M. VIOLLET-LE-DUC.

THE MILITARY ARCHITECTURE OF THE MIDDLE AGES, Translated from the French of M. VIOLLET-LE-DUC. By M. MACDERMOTT, Esq., Architect. With the 151 original French Engravings. Medium 8vo., cloth, price £1 1s.

EDITOR OF GLOSSARY.

SOME ACCOUNT OF DOMESTIC ARCHITECTURE IN ENGLAND, from Richard II. to Henry VIII. (or the Perpendicular Style). With Numerous Illustrations of Existing Remains from Original Drawings. By the EDITOR OF "THE GLOSSARY OF ARCHITECTURE." In 2 vols., 8vo., 1l. 10s.

Also,

VOL. I.—FROM WILLIAM I. TO EDWARD I. (or the Norman and Early English Styles). 8vo., 21s.

VOL. II.—FROM EDWARD I. TO RICHARD II. (the Edwardian Period, or the Decorated Style). 8vo., 21s.

The work complete, with 400 Engravings, and a General Index, 4 vols. 8vo., price £3 12s.

OUR ENGLISH HOME: its Early History and Progress. With Notes on the Introduction of Domestic Inventions. *Second Edition.* Crown 8vo., price 5s.

"It contains the annals of our English civilization, and all about our progress in social and domestic matters, how we came to be the family and people which we are. All this forms a book as interesting as a novel, and our domestic history is written not only with great research, but also with much spirit and liveliness."—*Christian Remembrancer.*

A NEW SERIES OF HISTORICAL TALES.

HISTORICAL TALES, *illustrating the chief events in Ecclesiastical History, British and Foreign, adapted for General Reading, Parochial Libraries, &c. In Monthly Volumes, with a Frontispiece, price 1s.*

This Series of Tales embraces the most important periods and transactions connected with the progress of the Church in ancient and modern times. They are written by authors of acknowledged merit, in a popular style, upon sound Church principles, and with a single eye to the inculcation of a true estimate of the circumstances to which they relate, and the bearing of those circumstances upon the history of the Church. By this means it is hoped that many, who now regard Church history with indifference, will be led to the perusal of its singularly interesting and instructive episodes.

Each Tale, although forming a link of the entire Series, is complete in itself.

Already published.

No. 1.—THE CAVE IN THE HILLS; or, Cæcilius Viriāthus.
No. 2.—THE EXILES OF THE CEBENNA. a Journal written during the Decian Persecution, by Aurelius Gratianus, Priest of the Church of Arles; and now done into English.
No. 3.—THE CHIEF'S DAUGHTER; or, The Settlers in Virginia.
No. 4.—THE LILY OF TIFLIS: a Sketch from Georgian Church History.
No. 5.—WILD SCENES AMONGST THE CELTS.
No 6.—THE LAZAR-HOUSE OF LEROS: a Tale of the Eastern Church in the Seventeenth Century.
No. 7.—THE RIVALS: a Tale of the Anglo-Saxon Church.
No. 8.—THE CONVERT OF MASSACHUSETTS.
No. 9.—THE QUAY OF THE DIOSCURI: a Tale of Nicene Times.
No. 10.—THE BLACK DANES.
No. 11.—THE CONVERSION OF ST. VLADIMIR; or, The Martyrs of Kief. A Tale of the Early Russian Church.
No. 12.—THE SEA-TIGERS: a Tale of Mediæval Nestorianism.
No. 13.—THE CROSS IN SWEDEN; or, The Days of King Ingi the Good.
No. 14.—THE ALLELUIA BATTLE; or, Pelagianism in Britain.
No. 15.—THE BRIDE OF RAMCUTTAH: A Tale of the Jesuit Missions to the East Indies in the Sixteenth Century.
No. 16.—ALICE OF FOBBING; or, The Times of Jack Straw and Wat Tyler.
No. 17.—THE NORTHERN LIGHT: a Tale of Iceland and Greenland in the Eleventh Century.
No. 18.—AUBREY DE L'ORNE; or, The Times of St. Anselm.
No. 19.—LUCIA'S MARRIAGE; or, The Lions of Wady-Araba.
No. 20.—WOLFINGHAM; or, The Convict-Settler of Jervis Bay: a Tale of the Church in Australia.
No. 21.—THE FORSAKEN; or, The Times of St. Dunstan.
No. 22.—THE DOVE OF TABENNA.—THE RESCUE: A Tale of the Moorish Conquest of Spain.
No. 23.—LARACHE: a Tale of the Portuguese Church in the Sixteenth Century.
No. 24.—WALTER THE ARMOURER; or, The Interdict: a Tale of the Times of King John.
No. 25.—THE CATECHUMENS OF THE COROMANDEL COAST.
No. 26.—THE DAUGHTERS OF POLA. Family Letters relating to the Persecution of Diocletian, now first translated from an Istrian MS.
No. 27.—AGNES MARTIN; or, The Fall of Cardinal Wolsey.
No. 28.—ROSE AND MINNIE; or, The Loyalists: a Tale of Canada in 1837.

NEW WORKS OF FICTION.

ALICE LISLE: A Tale of Puritan Times. Fcap. 8vo., cloth, 4s.

THE SCHOLAR AND THE TROOPER; OR, OXFORD DURING THE GREAT REBELLION. By the Rev. W. E. HEYGATE. *Second Edition.* Fcap. 8vo., cloth, 5s.

SOME YEARS AFTER: A Tale. Fcap. 8vo., cloth lettered, 7s.

ATHELINE; or, THE CASTLE BY THE SEA. A Tale. By LOUISA STEWART, Author of "Walks at Templecombe," "Floating away," &c. 2 vols., Fcap. 8vo. 9s.

MIGNONETTE: A SKETCH. By the Author of "The Curate of Holy Cross." 2 vols., Fcap., cloth, 10s.

THE CALIFORNIAN CRUSOE: A Tale of Mormonism. By the Rev. H. CASWALL, Vicar of Figheldean. Fcap. 8vo., with Illustration, cloth, 2s. 6d.

STORM AND SUNSHINE; OR, THE BOYHOOD OF HERBERT FALCONER. A Tale. By W. E. DICKSON, M.A., Author of "Our Workshop," &c. With Frontispiece, cloth, 2s.

AMY GRANT; OR, THE ONE MOTIVE. A Tale designed principally for the Teachers of the Children of the Poor. *Second Edition.* Fcap. 8vo., cloth, 3s. 6d.

THE TWO HOMES. A Tale. By the Author of "Amy Grant." *Third Edition.* Fcap. 8vo., cloth, 2s. 6d.

DAWN AND TWILIGHT. A Tale. By the Author of "Amy Grant," "Two Homes," &c. 2 vols. Fcap. 8vo., cloth, 7s.

KENNETH; OR, THE REAR-GUARD OF THE GRAND ARMY. By the Author of the "Heir of Redclyffe," "Heartsease," &c., &c. Fcap. 8vo., with Illustrations, 5s. *Fourth Edition.*

TALES FOR THE YOUNG MEN AND WOMEN OF ENGLAND. A Series of Tales adapted for Lending Libraries, Book Hawkers, &c. Fcap. 8vo., with Illustrations, strongly bound in coloured wrapper, 1s. each:—

No. 1. Mother and Son.
No. 2. The Recruit.
No. 3. The Strike.
No. 4. James Bright, the Shopman.
No. 5. Jonas Clint.
No. 6. The Sisters.
No. 7. Caroline Elton; or, Vanity and Jealousy. } 1s.
No. 8. Servants' Influence.
No. 9. The Railway Accident.
No. 10. Wanted, a Wife.
No. 11. Irrevocable.
No. 12. The Tenants at Tinkers' End.
No. 13. Windycote Hall.
No. 14. False Honour.
No. 15. Old Jarvis's Will.
No. 16. The Two Cottages.
No. 17. Squitch.
No. 18. The Politician.
No. 19. Two to One.
No. 20. Hobson's Choice. 6d.
No. 21. Susan. 4d.
No. 22. Mary Thomas; or, Dissent at Evenly. } 4d.

"To make boys learn to read, and then to place no good books within their reach, is to give them an appetite, and leave nothing in the pantry save unwholesome and poisonous food which, depend upon it, they will eat rather than starve."—*Sir W. Scott.*

NEW PAROCHIAL BOOKS.

CATECHETICAL WORKS, Designed to aid the Clergy in Public Catechising. Uniform in size and type with the "Parochial Tracts."

Already published in this Series.

I. CATECHETICAL LESSONS on the Creed. 6d.
II. CATECHETICAL LESSONS on the Lord's Prayer. 6d.
III. CATECHETICAL LESSONS on the Ten Commandments. 6d.
IV. CATECHETICAL LESSONS on the Sacraments. 6d.
V. CATECHETICAL LESSONS on the Parables of the New Testament. Part I. Parables I.—XXI. 1s.
VI. PART II. PARABLES XXII.—XXXVII. 1s.
VII. CATECHETICAL NOTES on the Thirty-Nine Articles. 1s. 6d.
VIII. CATECHETICAL LESSONS on the Order for Morning and Evening Prayer, and the Litany. 1s.
IX. CATECHETICAL LESSONS on the Miracles of our Lord. Part I. Miracles I.—XVII. 1s.
X. PART II. MIRACLES XVIII.—XXXVII. 1s.
XI. CATECHETICAL NOTES on the Saints' Days. 1s.
QUESTIONS ON THE COLLECTS, EPISTLES, AND GOSPELS, throughout the Year; edited by the Rev. T. L. CLAUGHTON, Vicar of Kidderminster. For the use of Teachers in Sunday-Schools. Two Parts, 18mo., cloth, each 2s. 6d.

COTTAGE PICTURES. Cottage Pictures from the Old Testament. Twenty-eight large Illustrations, coloured by hand. The set, folio, 7s. 6d.

COTTAGE PICTURES from the New Testament, (uniform with above). The set of Twenty-eight, 7s. 6d.

SCRIPTURE PRINTS FOR PAROCHIAL USE. Printed in Sepia, with Ornamental Borders. Price One Penny each; or the set in an ornamental envelope, One Shilling.

1. The Nativity.
2. St. John Preaching.
3. The Baptism of Christ.
4. Jacob's Dream.
5. The Transfiguration.
6. The Good Shepherd.
7. The Tribute-Money.
8. The Preparation for the Cross.
9. The Crucifixion.
10. Leading to Crucifixion.
11. Healing the Sick.
12. The Return of the Prodigal.

One hundred thousand have already been sold of these prints. They are also kept mounted and varnished, 3d. each.

TALES AND ALLEGORIES reprinted from the "PENNY POST."
Fcap. 8vo., with Illustrations.

FANNY DALE. 1s. 6d.
THE CHILD OF THE TEMPLE. 1s.
THE HEART-STONE. 10d.
FAIRTON VILLAGE. 8d.
GILL'S LAP. 8d.
FOOTPRINTS IN THE WILDERNESS. 6d.
TALES OF AN OLD CHURCH. 4d.
MARGARET OF CONWAY. 4d.
MARY WILBRAM. 4d.
MARION. 4d.

MARTINMAS. 4d.
OLD WINTERTON'S WILL. 4d.
MARY MERTON. 2d.
THE TWO WIDOWS. 2d.
LEFT BEHIND. 2d.
AMNEMON THE FORGETFUL. 2d.
EUSTATHES THE CONSTANT. 2d.

LITTLE TALES. 4d.
LITTLE ALLEGORIES. 2d.
LITTLE FABLES. 2d.

NEW AND STANDARD EDUCATIONAL WORKS. 13

THE FIFTH BOOK OF EUCLID.—The Propositions of the Fifth Book of Euclid proved Algebraically; with an Introduction, Notes, and Questions. By GEORGE STURTON WARD, M.A., Mathematical Lecturer in Magdalen Hall, and Public Examiner in the University of Oxford. Crown 8vo., price 2s. 6d.

Η ΚΑΙΝΗ ΔΙΑΘΗΚΗ. The Greek Testament with English Notes. By the Rev. EDWARD BURTON, D.D., sometime Regius Professor of Divinity in the University of Oxford. *Sixth Edition, with Index.* 8vo., cloth, 10s. 6d.

PASS AND CLASS. An Oxford Guide-Book through the Courses of *Literæ Humaniores*, Mathematics, Natural Science, and Law and Modern History. By MONTAGU BURROWS, M.A. *Second Edition, with some of the latest Examination Papers.* Fcap. 8vo., cloth, 5s.

ANNALS OF ENGLAND. An Epitome of English History. From Cotemporary Writers, the Rolls of Parliament, and other Public Records. 3 vols. Fcap. 8vo., with Illustrations, cloth, 15s. *Recommended by the Examiners in the School of Modern History at Oxford.*

Vol. I. From the Roman Era to the Death of Richard II. Cloth, 5s.
Vol. II. From the Accession of the House of Lancaster to Charles I. Cloth, 5s.
Vol. III. From the Commonwealth to the Death of Queen Anne. Cloth, 5s.

Each Volume is sold separately.

GRAMMARS AND DICTIONARIES.

JELF'S GREEK GRAMMAR.—A Grammar of the Greek Language, chiefly from the text of Raphael Kühner. By WM. EDW. JELF, M.A., Student of Ch. Ch. *Third Edition, greatly improved.* 2 vols. 8vo., 1l. 10s.

This Grammar is in general use at Oxford, Cambridge, Dublin, and Durham; at Eton, King's College, London, and most other public schools.

MADVIG'S LATIN GRAMMAR. A Latin Grammar for the Use of Schools. By Professor MADVIG, with additions by the Author. Translated by the Rev. G. WOODS, M.A. Uniform with JELF'S "Greek Grammar." *Fourth Edition.* 8vo., cloth, 12s.

Competent authorities pronounce this work to be the very best Latin Grammar yet published in England. This new Edition contains an Index to the Authors quoted.

LAWS OF THE GREEK ACCENTS. By JOHN GRIFFITHS, M.A. 16mo. *Eleventh Edition.* Price Sixpence.

Printed at the Oxford University Press.

A GREEK-ENGLISH LEXICON, based on the German work of F. Passow. By HENRY GEORGE LIDDELL, D.D., and ROBERT SCOTT, D.D. *Fifth Edition.* Crown 4to., cloth, £1 11s. 6d.

A LEXICON chiefly for the use of Schools, abridged from the Greek-English Lexicon of H. G. LIDDELL, D.D., and R. SCOTT, D.D. *Tenth Edition*, square 12mo., cloth, price 7s. 6d.

SCHELLER'S LATIN LEXICON. Translated by RIDDLE. *In sheets, £1 10s. (published at £4); a few copies may be had bound in calf, £2 nett.*

This is by far the most complete Latin Dictionary in existence, containing many hundred words more than any other.

SCAPULA. GREEK AND LATIN LEXICON. *In sheets*, 12s. *(published at £3 13s. 6d.); a few copies may be had bound, £1 1s. nett.*

Sold by J. H. and J. PARKER, Oxford, and 377, Strand, London;
and LONGMAN, GREEN, & Co., Paternoster-row.

OCTAVO EDITIONS OF CLASSICS.

THUCYDIDES, with Notes, chiefly Historical and Geographical. By the late T. ARNOLD, D.D. With Indices by the Rev. R. P. G. TIDDEMAN. *Fifth Edition.* 3 vols., 8vo., cloth lettered, £1 16s.

THE ETHICS OF ARISTOTLE. With Notes by the Rev. W. E. JELF, B.D., Author of "A Greek Grammar," &c. 8vo., cloth, 12s.
  The Text separately, 5s. The Notes separately, 7s. 6d.

SOPHOCLIS TRAGŒDIÆ, with Notes, adapted to the use of Schools and Universities. By THOMAS MITCHELL, M.A. 2 vols. 8vo., £1 8s.

The following Plays may also be had separately, at 5s. each:—

PHILOCTETES.	AJAX.
TRACHINIÆ.	ELECTRA.

ŒDIPUS COLONEUS.

A SERIES OF GREEK AND LATIN CLASSICS
FOR THE USE OF SCHOOLS.

GREEK AUTHORS.

	Paper.		Bound.	
	s.	d.	s.	d.
Æschylus	2	6	3	0
Aristophanes. 2 vols.	5	0	6	0
Euripides. 3 vols.	5	0	6	6
———— Tragœdiæ Sex	3	0	3	6
Sophocles	2	6	3	0
Homeri Ilias	3	0	3	6
———— Odyssea	2	6	3	0
Aristotelis Ethica	1	6	2	0
Demosthenes de Corona, et Æschines in Ctesiphontem	1	6	2	0
Herodotus. 2 vols.	5	0	6	0
Thucydides. 2 vols.	4	0	5	0
Xenophontis Memorabilia	1	0	1	4
———— Anabasis	1	6	2	0

LATIN AUTHORS.

	Paper.		Bound.	
Horatius	1	6	2	0
Juvenalis et Persius	1	0	1	6
Lucanus	2	0	2	6
Lucretius	1	6	2	0
Phædrus	1	0	1	4
Virgilius	2	0	2	6
Cæsaris Commentarii, cum Supplementis Auli Hirtii et aliorum	2	0	2	6
———— Commentarii de Bello Gallico	1	0	1	6
Cicero De Officiis, de Senectute, et de Amicitia	1	6	2	0
Ciceronis Tusculanarum Disputationum Libri V.	1	6	2	0
———— Orationes Selectæ, *in the press*				
Cornelius Nepos	1	0	1	4
Livius. 4 vols.	5	0	6	0
Sallustius	1	6	2	0
Tacitus. 2 vols.	4	0	5	0

NEW SERIES OF ENGLISH NOTES.

Pocket Editions of the following have been published with Short Notes.

		s.	d.
SOPHOCLES	Ajax (*Text and Notes*)	1	0
	Electra ,,	1	0
	Œdipus Rex ,,	1	0
	Œdipus Coloneus ,,	1	0
	Antigone ,,	1	0
	Philoctetes ,,	1	0
	Trachiniæ ,,	1	0
	The Notes only, in one vol., cloth, 3s.		
ÆSCHYLUS	Prometheus Vinctus (*Text and Notes*)	1	0
	Septem Contra Thebas ,,	1	0
	Persæ ,,	1	0
	Agamemnon ,,	1	0
	Choephoræ ,,	1	0
	Eumenides ,,	1	0
	Supplices ,,	1	0
	The Notes only, in one vol., cloth, 3s. 6d.		
EURIPIDES	Hecuba (*Text and Notes*)	1	0
	Medea ,,	1	0
	Orestes ,,	1	0
	Hippolytus ,,	1	0
	Phœnissæ ,,	1	0
	Alcestis ,,	1	0
	The Notes only, in one vol., cloth, 3s.		
ARISTOPHANES	The Knights, (*Text and Notes*)	1	0
	Acharnians ,,	1	0
DEMOSTHENES	De Corona, (*Text and Notes*,)	2	0
OLYNTHIAC ORATIONS, *in the press.*			
ÆSCHINES	In Ctesiphontem, (*Text and Notes*)	2	0
HOMERUS	Ilias, Lib. i.—vi. (*Text and Notes*,)	2	0
VIRGILIUS	Bucolica, (*Text and Notes*,)	1	0
	Georgica ,,	2	0
	Æneidos, Lib. i.—iii. ,,	1	0
HORATIUS	Carmina, &c., (*Text and Notes*,)	2	0
	Satiræ ,,	1	0
	Epistolæ et Ars Poetica ,,	1	0
	The Notes only, in one vol., cloth, 2s.		
SALLUSTIUS	Jugurtha, (*Text and Notes*,)	1	6
	Catilina ,,	1	0
M. T. CICERO	Pro Milone, (*Text and Notes*)	1	0
	In Catilinam ,,	1	0
	Pro Lege Manilia, and Pro Archia ,,	1	0
	De Senectute and De Amicitia ,,	1	0
LIVIUS	Lib. xxi.—xxiv. (*Text and Notes*,)	4	0
CÆSAR	Lib. i.—iii. (*Text and Notes*,)	1	0
CORNELIUS NEPOS, (*Text and Notes*,)		1	6
PHÆDRUS	Fabulæ, (*Text and Notes*,)	1	0

Other portions of several of the above-named Authors are in preparation.

"The notes contain sufficient information, without affording the pupil so much assistance as to supersede all exertion on his part."—*Athenæum*, Jan. 27, 1855.

"Be all this as it may, it is a real benefit to public schoolboys to be able to purchase any Greek Play they want for One Shilling. When we were introduced to Greek Plays, about forty years ago, we had put into our hands a portly 8vo. volume, containing Porson's four Plays, without one word of English in the shape of notes; and we have no doubt the book cost nearer twenty than ten shillings, and after all was nothing near so useful as these neat little copies at One Shilling each."—*Educational Times.*

THE GENTLEMAN'S MAGAZINE.—As time has passed on, the Magazine which SYLVANUS URBAN, now one hundred and thirty-two years ago commenced, has given rise to very many others. It is true that the greater part have died soon after their birth, but still there are some occupying a portion of the ground once covered by SYLVANUS URBAN: this has rendered it necessary for him to prescribe certain limits for his labours. While Magazine after Magazine has been set on foot, none has ever clearly marked out for itself the ground which has ever held the most prominent place in the GENTLEMAN'S MAGAZINE, namely, History and Archæology. The GENTLEMAN'S MAGAZINE has done more than any periodical to support and promote archæological tastes and studies. In former years it is true much of its space was occupied with general literature, but of late, as its ground became more circumscribed, the Magazine was able to give a more complete *résumé* of archæological progress and labours. It therefore depended from that time more especially upon archæological and historical students to supply the place of those supporters whom death year by year removes.

An independent organ in such a study as archæology is of the greatest importance, and he therefore appeals to the several Societies to receive him as such, believing that his Magazine, by shewing what other Societies are doing, stimulates the members of each Society to further exertion; that by its independence it prevents the injury that is often done to a science by that narrowness of view which small separate Societies tend to engender; and finally, that by treating of various subjects distinct from those coming beneath the scope of such Societies, he supplements their labours without interfering with them.

Published Monthly, with numerous Illustrations, 8vo., 2s. 6d.

All Communications to be addressed to Mr. URBAN, 377, STRAND, W.C.

THE PENNY POST. A Church of England Illustrated Magazine, issued Monthly. Price One Penny.

ENLARGEMENT OF THE PENNY POST.

Each number of 1863 will consist of Thirty-two Pages, and numerous Illustrations, containing Tales, Stories, Allegories; Notes on Religious Events of the Day; Essays, Doctrinal and Practical. The object is to combine amusement with instruction; to provide healthy and *interesting* reading adapted for the Village as well as the Town. A part of each number is devoted to the "Children's Corner." The Editor's Box is continued.

MONTHLY—ONE PENNY.

Subscribers' names received by all Booksellers and Newsmen.

PARKER'S CHURCH CALENDAR FOR 1863, price 6d., being the Eighth Year of Issue.

CONTENTS:—THE CALENDAR, with the Daily Lessons.—THE CHURCH: Statistics of the Church in England, Ireland, Scotland, the Colonies, &c. CONVOCATION: Province of Canterbury; Province of York. The Archbishops and Bishops of ENGLAND; Deans; Archdeacons; Proctors of Convocation; Benefices; Church and other Sittings, &c. The Archbishops and Bishops in IRELAND, SCOTLAND, and the COLONIES; and other information. The Bishops in AMERICA, with Statistics of the Dioceses, Extent, Population, Clergy, &c., &c. : First Bishops of the American Church; Statistical Summaries.—THE UNIVERSITIES, &c.: The University of OXFORD; the University of CAMBRIDGE; and other Universities. Theological Colleges; Training Institutions; Public Schools. Societies and Institutions; Church Unions; Colleges.—THE STATE, &c.: The State and Royal Family. Members of the House of Peers and Commons. The Kings and Queens of England. Statistics of the Population. Postal Regulations, and a variety of other information.

Crown 8vo., price 1s.; 140 pages of close type, in wrapper,

THE OXFORD DIOCESAN CALENDAR, AND CLERGY LIST for 1863. Issued under the sanction of the Lord Bishop of the Diocese.

Also,

THE LONDON DIOCESAN CALENDAR, AND CLERGY LIST for 1863.

A Diary is also printed, uniform with the Calendar.
The two together, bound in roan, 3s.

Oxford and London: J. H. and J. PARKER.

www.ingramcontent.com/pod-product-compliance
Lightning Source LLC
Chambersburg PA
CBHW020247170426
43202CB00008B/263